AF659173

Electromagnetism

Electromagnetism

Hillary D. Brewster

Oxford Book Company
Jaipur ▪ New Delhi

Published By

Oxford Book Company

Regd. Off. "Bony Residency", Gate No. 2
Opp. Tilak Public School, Vishweswariya Nagar,
Gopalpura Road, Jaipur - 302018, India
Ph: 0141 - 2761280, Telefax: 2761381
email - oxfordbookjpr@gmail.com
website - oxfordbookcompany.com

Delhi Office: 102, 1st Floor, Satyam House,
4327/3, Ansari Road, Darya Ganj,
New Delhi - 110002, India
Ph: 011-45652440

Published in India, 2025

© All Rights Reserved

ISBN : 9789380179339

Price : ₹ 2150/-

All Rights Reserved. No part of this publication may be reproduced, stored in a retrieval system, or transmitted, in any form or by any means, electronic, mechanical, photocopying, recording, scanning or otherwise, without the prior written permission of the copyright owner.

Typeset by Shivangi Computers, Jaipur

Printed and bound in India by Thomson Press Ltd.,
New Delhi

Preface

The book "Electromagnetism" is prepared keeping in mind the languages used for science students as well as for those who are interested to understand better about Electromagnetism. Electromagnetism is the physics of the electromagnetic field - a field which exerts a force on particles that possess the property of electric charge, and is in turn affected by the presence and motion of those particles. A changing electromagnet produces an electric field. This is the phenomenon of electromagnetic induction, the basis of operation for electrical generators, induction motors, and transformers.

Similarly, a changing electric field generates a magnetic field. Because of this interdependence of the electric and magnetic fields, it makes sense to consider them as a single coherent entity - the electromagnetic field. An accurate theory of electromagnetism, known as classical electromagnetism, was developed by various physicists over the course of the 19th century, culminating in the work of James Clerk Maxwell, who unified the preceding developments into a single theory and discovered the electromagnetic nature of light.

In classical electromagnetism, the electromagnetic field obeys a set of equations known as Maxwell's equations, and the electromagnetic force is given by the Lorentz force law. To understand these theories and models, as well as to generate new ones, the student will find it helpful to be familiar with the topics discussed in this book.

Hilary. D. Brewster

Contents

Preface *v*

1. Introduction 1
2. Magnetism and its Properties 37
3. Electromagnetic Waves 134
4. The Universe and the Gravity 150
5. Faraday's Unified Field Concept 164
6. Einstein's Unified Field Concept 173
7. The Mach Principle 184
8. Inertial Mass 199
9. Classical Electromagnetism 230

Index *275*

Chapter 1

Introduction

While it is never safe to affirm that the future of the Physical Sciences has no marvels in store even more astonishing than those of the past, it seems probable that most of the grand underlying principles have been firmly established and that further advances are to be sought chiefly in the rigorous application of these principles to all the phenomena which come under our notice. ... An eminent physicist has remarked that the future truths of Physical Science are to be looked for in the sixth place of decimals.

Electromagnetism is the physics of the electromagnetic field: a field which exerts a force on particles that possess the property of electric charge, and is in turn affected by the presence and motion of those particles. A changing electromagnet produces an electric field(this is the phenomenon of electromagnetic induction, the basis of operation for electrical generators, induction motors, and transformers). Similarly, a changing electric field generates a magnetic field. Because of this interdependence of the electric and magnetic fields, it makes sense to consider them as a single coherent entity - the electromagnetic field.

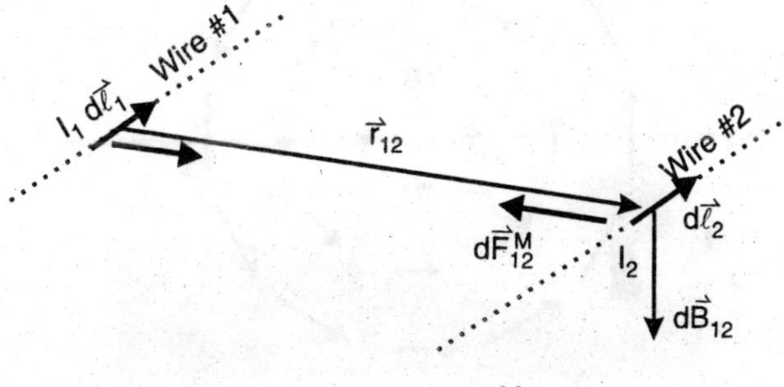

Fig. The Magnetic Force $d\vec{F}_{12}^M$ on Current Element $I_2 d\vec{l}_2$ Due to Current Element $I_1 d\vec{l}_1$.

The magnetic field is produced by the motion of electric charges, i.e., electric current. The magnetic field causes the magnetic force associated with magnets. The theoretical implications of electromagnetism led to the development of special relativity by Hermann Minkowski and Albert Einstein in 1905.

While preparing for an evening lecture on 21 April 1820, Hans Christian Orsted developed an experiment which provided evidence that surprised him.

As he was setting up his materials, he noticed a compass needle deflected from magnetic north when the electric current from the battery he was using was switched on and off.

This deflection convinced him that magnetic fields radiate from all sides of a wire carrying an electric current, just as light and heat do, and that it confirmed a direct relationship between electricity and magnetism. At the time of discovery, Orsted did not suggest any satisfactory explanation of the phenomenon, nor did he try to represent the phenomenon in a mathematical framework.

However, three months later he began more intensive investigations. Soon thereafter he published his findings, proving that an electric current produces a magnetic field as it flows through a wire. The CGS unit of magnetic induction(oersted) is named in honour of his contributions to the field of electromagnetism.

His findings resulted in intensive research throughout the scientific community in electrodynamics. They influenced French physicist André-Marie Ampère's developments of a single mathematical form to represent the magnetic forces between current-carrying conductors. Orsted's discovery also represented a major step toward a unified concept of energy.

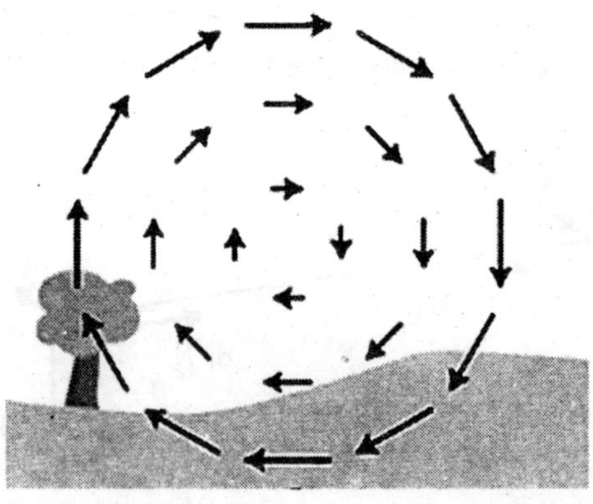

Fig. The Moving Charge Responsible for this Magnetic Field.

Orsted was not the first person to examine the relation between electricity and magnetism. In 1802 Gian Domenico Romagnosi, an Italian legal scholar, deflected a magnetic needle by electrostatic charges. He interpreted his observations as *The Relation* between electricity and magnetism. Actually, no galvanic current existed in the setup and hence no electromagnetism was present.

An account of the discovery was published in 1802 in an Italian newspaper, but it was largely overlooked by the contemporary scientific community.

This unification, which was observed by Michael Faraday, extended by James Clerk Maxwell, and partially reformulated by Oliver Heaviside and Heinrich Hertz, is one of the accomplishments of 19th century mathematical physics.

It had far-reaching consequences, one of which was the understanding of the nature of light.

As it turns out, what is thought of as "light" is actually a propagating oscillatory disturbance in the electromagnetic field, i.e., an electromagnetic wave.

Different frequencies of oscillation give rise to the different forms of electromagnetic radiation, from radio waves at the lowest frequencies, to visible light at intermediate frequencies, to gamma rays at the highest frequencies.

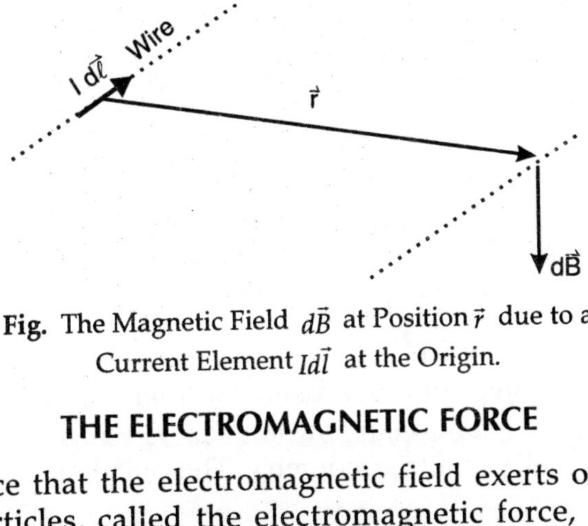

Fig. The Magnetic Field $d\vec{B}$ at Position $\vec{r}$ due to a Current Element $Id\vec{l}$ at the Origin.

THE ELECTROMAGNETIC FORCE

The force that the electromagnetic field exerts on electrically charged particles, called the electromagnetic force, is one of the fundamental forces, and is responsible for most of the forces we experience in our daily lives. The other fundamental forces are the strong nuclear force(which holds atomic nuclei together), the weak nuclear force and the gravitational force. All other forces are ultimately derived from these fundamental forces. The electromagnetic force is the one responsible for practically all the phenomena encountered in daily life, with the exception of gravity.

All the forces involved in interactions between atoms can be traced to the electromagnetic force acting on the electrically charged protons and electrons inside the atoms. This includes the forces we experience in "pushing" or "pulling" ordinary material objects, which come from the intermolecular forces between the individual molecules in our bodies and those in the objects. It also includes all forms of chemical phenomena, which arise from interactions between electron orbitals. There is a saying that in computer science, there are only three nice numbers: zero, one, and however many you please.

In other words, computer software shouldn't have arbitrary limitations like a maximum of 16 open files, or 256 e-mail messages per mailbox. When superposing the fields of long, straight wires, the really interesting cases are one wire, two wires, and infinitely many wires. With an infinite number of wires, each carrying an infinitesimal current, we can create sheets of current. Such a sheet has a certain amount of current per unit length, which we notate ç(Greek letter eta). For the y component, we have

$$
\begin{aligned}
B_y &= \int \frac{2k\,dI}{c^2 R}\cos\theta \\
&= \int_{-a}^{b} \frac{2k\eta\,dy}{c^2 R}\cos\theta \\
&= \frac{2k\eta}{c^2}\int_{-a}^{b} \frac{\cos\theta}{R}\,dy \\
&= \frac{2k\eta}{c^2}\int_{-a}^{b} \frac{z\,dy}{y^2+z^2} \\
&= \frac{2k\eta}{c^2}\left(\tan^{-1}\frac{b}{z} - \tan^{-1}\frac{-a}{z}\right) \\
&\frac{2k\eta\Upsilon}{c^2},
\end{aligned}
$$

where in the last step we have used the identity $\tan^{-1}(-x)=-\tan^{-1}x$, combined with the relation $\tan^{-1}b/z+\tan^{-1}a/z=\Upsilon$, which can be verified with a little geometry and trigonometry. The calculation of B_z is left as an exercise.

More interesting is what happens underneath the sheet: by the right-hand rule, all the currents make rightward contributions to the field there, so B_y abruptly reverses itself as we pass through the sheet.

Close to the sheet, the angle Υ approaches π, so we have

$$B_y = \frac{2\pi k\eta}{c^2}$$

In one case the sources are charges and the field is electric; in the other case we have currents and magnetic fields. In both cases we find that the field changes suddenly when we pass through a sheet of sources, and the amount of this change doesn't depend on the size of the sheet. It was this type of reasoning that eventually led us to Gauss' law in the case of electricity, we will see that a similar approach can be used with magnetism.

The difference is that, whereas Gauss' law involves the flux, a measure of how much the field *spreads out*, the corresponding law for magnetism will measure how much the field *curls*.

We've already seen that what one observer perceives as an electric field, another observer may perceive as a magnetic field.

An observer flying along above a charged sheet will say that the charges are in motion, and will therefore say that it is both a sheet of current and a sheet of charge.

Instead of a pure electric field, this observer will experience a combination of an electric field and a magnetic one.

Energy in the Magnetic Field

The energy density of the magnetic field must be proportional to $|B|^2$, which we can write as B^2 for convenience.

To pin down the constant of proportionality, we now need to do: find one example where we can calculate the mechanical work done by the magnetic field, and equate that to the amount of energy lost by the field itself. The easiest example is two parallel sheets of charge, with their currents in opposite directions.

$$dU_m = \frac{c^2}{8\pi k} B^2 dv.$$

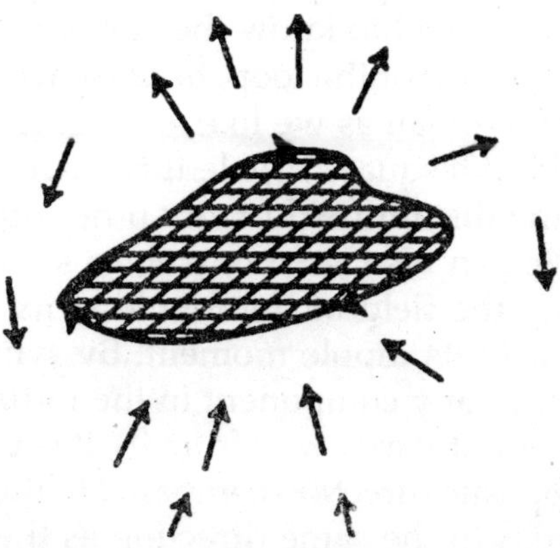

Fig. The Field of any Planar Current Loop can be Found by Breaking it down into Square Dipoles.

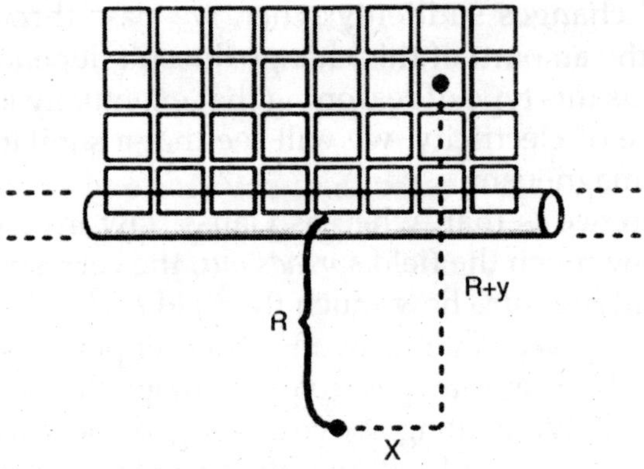

Fig. Setting up the Integral.

SUPERPOSITION OF DIPOLES

The Distant Field of a Dipole, in its Midplane

Most current distributions cannot be broken down into long, straight wires, and subsection has exhausted most of the interesting cases we can handle in this way. A much more useful building block is a square current loop.

We have already seen how the dipole moment of an irregular current loop can be found by breaking the loop down into square dipoles because the currents in adjoining squares cancel out on their shared edges. Likewise if we could find the magnetic field of a square dipole, then we could find the field of any planar loop of current by adding the contributions to the field from all the squares.

The field of a square-loop dipole is very complicated close up, but luckily for us, we only need to know the current at distances that are large compared to the size of the loop, because we're free to make the squares on our grid as small as we like.

The *distant* field of a square dipole turns out to be simple, and is no different from the distant field of any other dipole with the same dipole moment. We can also save ourselves some work if we only worry about finding the field of the dipole in its own plane, i.e. the plane perpendicular to its dipole moment. By symmetry, the field in this plane cannot have any component in the radial direction(inward toward the dipole, or outward away from it); it is perpendicular to the plane, and in the opposite direction compared to the dipole vector.(The field *inside* the loop is in the same direction as the dipole vector, but we're interested in the distant field.) Letting the dipole vector be along the z axis, we find that the field in the x-y plane is of the form $B_z=f(r)$, where $f(r)$ is some function that depends only on r, the distance from the dipole.

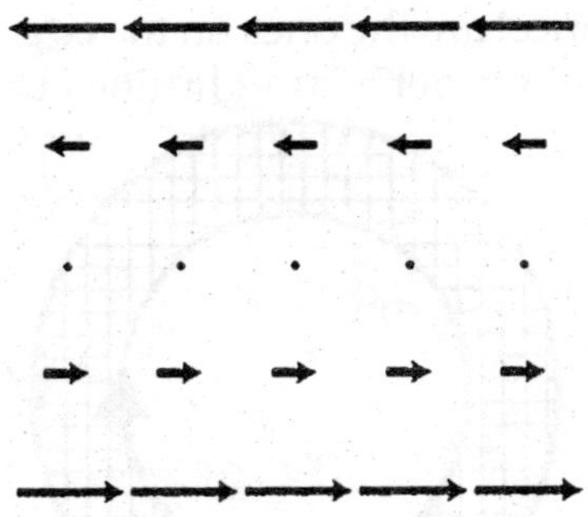

Fig. The Field $-y\hat{x}$.

We can pin down the result even more without any math. We know that the magnetic field made by a current always contains a factor of k/c^2, which is the coupling constant for magnetism.

We also know that the field must be proportional to the dipole moment, $m=IA$. Fields are always directly proportional to currents, and the proportionality to area follows because dipoles add according to their area.

For instance, a square dipole that is 2 micrometers by 2 micrometers in size can be cut up into four dipoles that are 1 micrometer on a side. This tells us that our result must be of the form $B_z=(k/c^2)(IA)g(r)$. Now if we multiply the quantity$(k/c^2)(IA)$ by the function $g(r)$, we have to get units of teslas, and this only works out if $g(r)$ has units of m^{-3}, so our result must be of the form

$$B_z = \frac{\beta kIA}{c^2 r^3},$$

where β is a unitless constant. Thus our only task is to determine β, and we will have determined the field of the dipole(in the plane of its current, i.e., the midplane with respect to its dipole moment vector).

If we wanted to, we could simply build a dipole, measure its field, and determine β empirically.

Better yet, we can get an exact result if we take a current loop whose field we know exactly, break it down into infinitesimally small squares, and integrate to find the total field, set this result equal to the known expression for the field of the loop, and solve for β.

There's just one problem here. We don't yet know an expression for the field of *any* current loop of *any* shape — all we know is the field of a long, straight wire.

Are we out of luck? No, because, we can make a long, straight wire by putting together square dipoles! Any square dipole away from the edge has all four of its currents canceled by its neighbors. The only currents that don't cancel are the ones on the edge, so by superimposing all the square dipoles, we get a straight-line current.

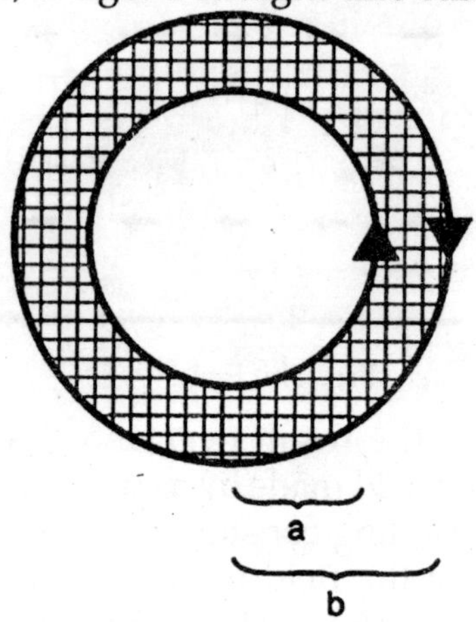

Fig. Counter Rotating Currents

This might seem strange. If the squares on the interior have all their currents canceled out by their neighbors, why do we even need them? Well, we need the squares on the edge in order to make the straight-line current.

We need the second row of squares to cancel out the currents at the top of the first row of squares, and so on.

Integrating, we have

$$B_z = \int_{y=0}^{\infty} \int_{x=-\infty}^{\infty} dB_z,$$

where dB_z is the contribution to the total magnetic field at out point of interest, which lies a distance R from lies a distance R from the wire.

$$\begin{aligned} B_z &= \int_{y=0}^{\infty} \int_{z=-\infty}^{\infty} \frac{\beta kI\, dA}{c^2 r^3} \\ &= \frac{\beta kI}{c^2} \int_{y=0}^{\infty} \int_{x=-\infty}^{\infty} \frac{1}{\left[x^2+(R+y)^2\right]^{-3/2}} dx\, dy \\ &= \frac{\beta kI}{c^2 R^3} \int_{y=0}^{\infty} \int_{x=-\infty}^{\infty} \left[\left(\frac{x}{R}\right)^2 + \left(1+\frac{y}{R}\right)^2\right]^{-3/2} dx\, dy \end{aligned}$$

This can be simplified with the substitutions

$$x = Ru,\ y = Rv, \text{ and } dx\ dy = R^2\ du\ dv:$$

$$B_z = \frac{\beta kI}{c^2 R}\int_{v=0}^{\infty}\int_{u=-\infty}^{\infty}\frac{1}{\left[u^2+(1+v)^2\right]^{3/2}}\,du\,dv$$

The u integral is of the form $\int_{-\infty}^{\infty}(u^2+b)^{-3/2}du = 2/b^2$,

so $B_z = \frac{\beta kI}{c^2 R}\int_{v=0}^{\infty}\frac{1}{(1+v)^2}dv$, and the remaining v integral is equals 2,

so $B_z = \frac{2\beta kI}{c^2 R}$.

This is the field of a wire, which we already know equals $2kI/c^2R$, so we have β=1. Remember, the point of this whole calculation was not to find the field of a *wire*, which we already knew, but to find the unitless constant â in the expression for the field of a *dipole*. The distant field of a dipole, in its midplane, is therefore B_z=β kIA/c^2r^3= kIA/c^2r^3, or, in terms of the dipole moment, $_z = \frac{km}{c^2 r^3}$.

Out of its Midplane

Let's compare with an electric dipole. An electric dipole, unlike a magnetic one, can be built out of two opposite monopoles, i.e., charges, separated by a certain distance, and it is then straightforward to show by vector addition that the field of an electric dipole is

$$E_z = kD(3\cos^2\theta - 1)r^{-3}$$
$$E_R = kD(3\sin\theta\cos\theta)r^{-3},$$

where r is the distance from the dipole to the point of interest, è is the angle between the dipole vector and the line connecting the dipole to this point, and E_z and E_R are, respectively, the components of the field parallel to and perpendicular to the dipole vector.

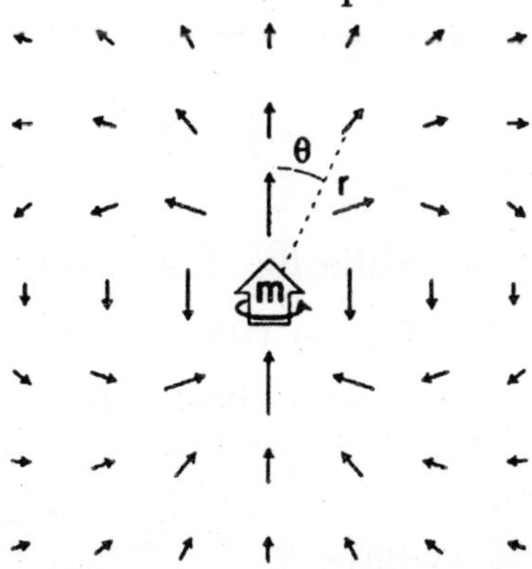

Fig. The Field of a Dipole.

In the midplane, θ equals π/2, which produces E_z=$-kDr^{-3}$ and E_R=0. This is the same as the field of a *magnetic* dipole in its midplane, except that the electric coupling constant k replaces the magnetic version k/c^2, and the electric dipole moment D is substituted for the magnetic dipole moment m.

It is therefore reasonable to conjecture that by using the same presto-change-o recipe we can find the field of a magnetic dipole outside its midplane:

$$B_z = \frac{km}{c^2}\left(3\cos^2\theta - 1\right)r^{-3}$$

$$B_R = \frac{km}{c^2}(3\sin\theta\cos\theta)r^{-3}.$$

This turns out to be correct.

Concentric, Counterrotating Currents

Two concentric circular current loops, with radii a and b, carry the same amount of current I, but in opposite directions.

We can produce these currents by tiling the region between the circles with square current loops, whose currents all cancel each other except at the inner and outer edges.

The flavour of the calculation is the same as the one in which we made a line of current by filling a half-plane with square loops. The main difference is that this geometry has a different symmetry, so it will make more sense to use polar coordinates instead of x and y. The field at the centre is

$$\begin{aligned} B_z &= \int \frac{kI}{c^2 r^3} dA \\ &= \int_{r=a}^{b} \frac{kI}{c^2 r^3} \cdot 2\pi r dr \\ &= \frac{2\pi kI}{c^2}\left(\frac{1}{a} - \frac{1}{b}\right). \end{aligned}$$

The positive sign indicates that the field is out of the page.

Field at the Centre of a Circular Loop

What is the magnetic field at the centre of a circular current loop of radius a, which carries a current I? Taking the limit of that result as b approaches infinity, we have $B_z = \frac{2\pi kI}{c^2 a}$

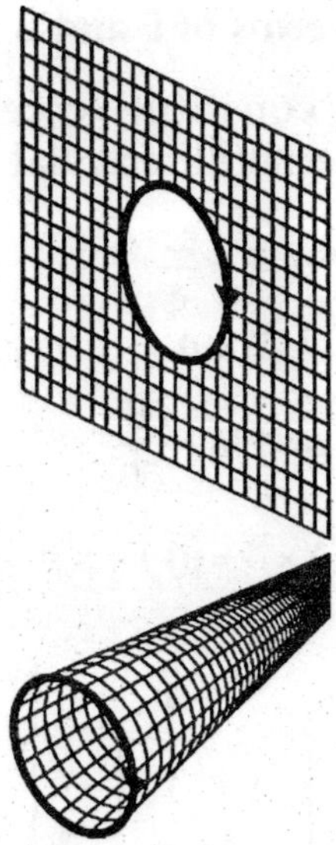

Fig. Two Ways of Making a Current Loop out of Square Dipoles.

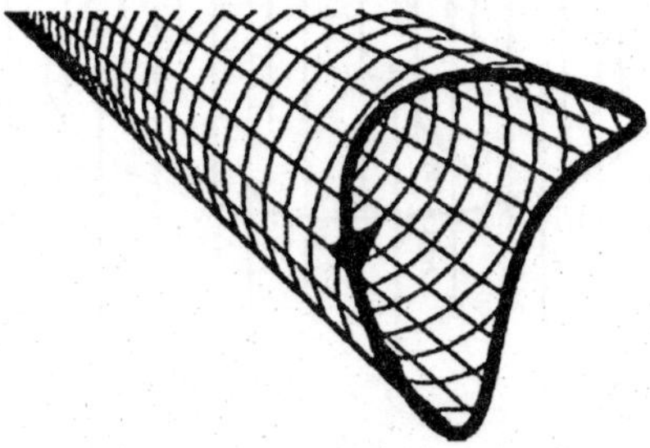

Fig. The New Method Can Handle Non-Planar Currents.

CLASSICAL ELECTRODYNAMICS

The scientist William Gilbert proposed, in his *De Magnete*(1600), that electricity and magnetism, while both capable of causing attraction and repulsion of objects, were distinct effects. Mariners had noticed that lightning strikes had the ability to disturb a compass needle, but the link between lightning and electricity was not confirmed until Benjamin Franklin's proposed experiments in 1752. One of the first to discover and publish a link between man-made electric current and magnetism was Romagnosi, who in 1802 noticed that connecting a wire across a voltaic pile deflected a nearby compass needle. However, the effect did not become widely known until 1820, when Orsted performed a similar experiment. Orsted's work influenced Ampère to produce a theory of electromagnetism that set the subject on a mathematical foundation.

Maxwell's Equations in Terms of E and B for Linear Materials

Substituting in the constitutive relations above, Maxwell's equations in a linear material(differential form only) are:

$$\Delta \cdot E = \frac{\rho f}{\in}$$

$$\Delta \cdot B = 0$$

$$\Delta \times E = -\frac{\partial B}{\partial t}$$

$$\Delta \times B = \mu J f + \mu \in \frac{\partial E}{\partial t}.$$

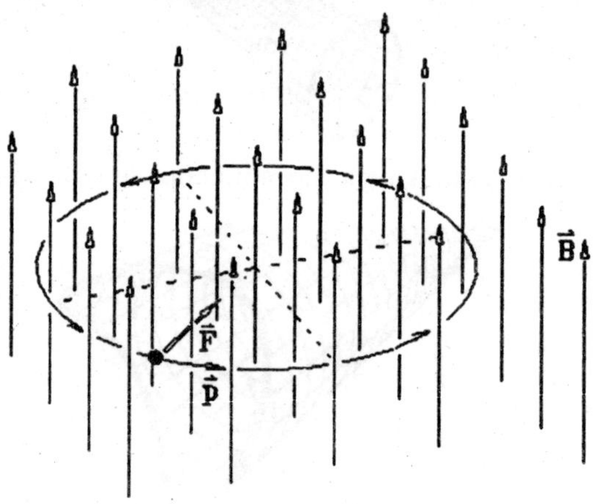

Fig. Path of a Charged Particle with Momentum $\vec{p}$ in a Uniform, Static Magnetic Field $\vec{B}$ Perpendicular to $\vec{p}$.

An accurate theory of electromagnetism, known as classical electromagnetism was developed by various physicists over the course of the 19th century, culminating in the work of James Clerk Maxwell, who unified the preceding developments into a single theory and discovered the electromagnetic nature of light.

In classical electromagnetism, the electromagnetic field obeys a set of equations known as Maxwell's equations, and the electromagnetic force is given by the Lorentz force law. One of the peculiarities of classical electromagnetism is that it is difficult to reconcile with classical mechanics, but it is compatible with special relativity. According to Maxwell's equations, the speed of light in a vacuum is a universal constant, dependent only on the electrical permittivity and magnetic permeability of free space.

This violates Galilean invariance, a long-standing cornerstone of classical mechanics. One way to reconcile the two theories is to assume the existence of a luminiferous aether through which the light propagates. However, subsequent experimental efforts failed to detect the presence of the aether.

After important contributions of Hendrik Lorentz and Henri Poincaré, in 1905, Albert Einstein solved the problem with the introduction of special relativity, which replaces classical kinematics with a new theory of kinematics that is compatible with classical electromagnetism. In addition, relativity theory shows that in moving frames of reference a magnetic field transforms to a field with a nonzero electric component and vice versa; thus firmly showing that they are two sides of the same coin, and thus the term "electromagnetism".

The Photoelectric Effect

In another paper published in that same year, Albert Einstein undermined the very foundations of classical electromagnetism. His theory of the photoelectric effect(for which he won the Nobel prize for physics) posited that light could exist in discrete particle-like quantities, which later came to be known as photons.

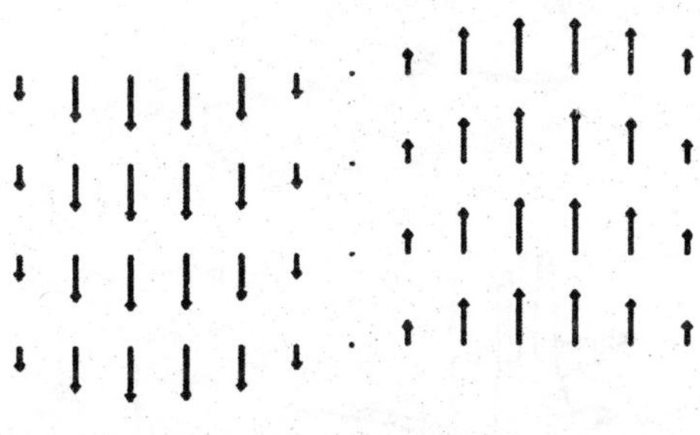

Fig. A Magnetic Field in the Form of a Sine Wave.

Einstein's theory of the photoelectric effect extended the insights that appeared in the solution of the ultraviolet catastrophe presented by Max Planck in 1900.

In his work, Planck showed that hot objects emit electromagnetic radiation in discrete packets, which leads to a finite total energy emitted as black body radiation.

Both of these results were in direct contradiction with the classical view of light as a continuous wave. Planck's and Einstein's theories were progenitors of quantum mechanics, which, when formulated in 1925, necessitated the invention of a quantum theory of electromagnetism. This theory, completed in the 1940s, is known as quantum electrodynamics(or "QED"), and is one of the most accurate theories known to physics. The term electrodynamics is sometimes used to refer to the combination of electromagnetism with mechanics, and deals with the effects of the electromagnetic field on the dynamic behaviour of electrically charged particles.

Units

Electromagnetic units are part of a system of electrical units based primarily upon the magnetic properties of electric currents, the fundamental cgs unit being the ampere. The units are:

- Ampere(current)
- Coulomb(charge)
- Farad(capacitance)
- Henry(inductance)
- Ohm(resistance)
- Volt(electric potential)
- Watt(power)
- Tesla(magnetic field)

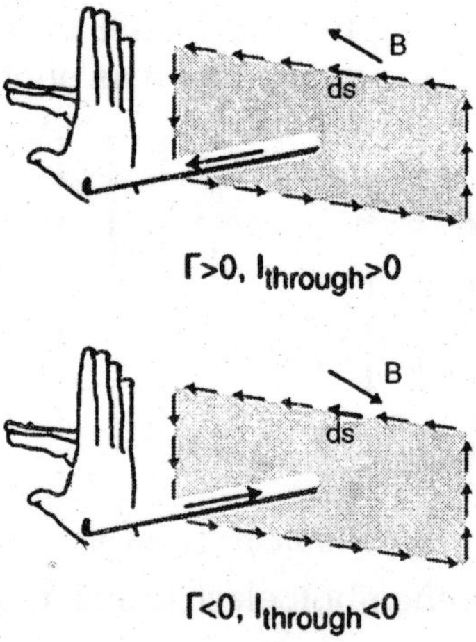

Fig. Positive and Negative Signs in Ampère's Law.

In the electromagnetic cgs system, electrical current is a fundamental quantity defined via Ampère's law and takes the permeability as a dimensionless quantity(relative permeability) whose value in a vacuum is unity.

As a consequence, the square of the speed of light appears explicitly in some of the equations interrelating quantities in this system.

ELECTROMAGNETIC PHENOMENA

In the theory, electromagnetism is the basis for optical phenomena, as discovered by James Clerk Maxwell while he studied electromagnetic waves. Light, being an electromagnetic wave, has properties that can be explained through Maxwell's equations, such as reflection, refraction, diffraction, interference and others. Relativity is born on the electromagnetic fields, as shown by Albert Einstein when he tried to make the electromagnetic theory compatible with Planck's radiation formula.

This catalog description was almost surely written by Albert A. Michelson, who was then head of the physics department and who had spoken very nearly the same words in a convocation address in 1894. The eminent gentleman whom he quotes may well have been Lord Kelvin. That 1894 talk proved to be well timed for contradiction. In quick succession, beginning soon afterward, there came the discovery of X-rays, radioactivity, the electron, special relativity, and the beginnings of quantum mechanics—all of this within a decade centered around the turn of the century.

Indeed, it was Michelson himself, working together with E. W. Morley, who in 1881 had carried out the crucial experiment that was later recognized as a foundation stone of special relativity. Both Michelson and Kelvin received Nobel Prize awards in the early years of the twentieth century.

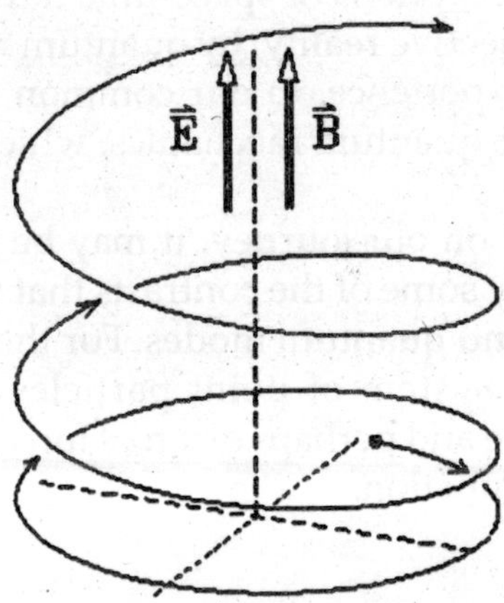

Fig. Path of a Charged Particle in Parallel $\vec{E}$ and $\vec{B}$ Fields.

In short, all the grand underlying principles had *not* been firmly established by the end of the nineteenth century. This cautionary tale should not be told with any sense of mockery. Those distinguished scientists—and there were others who spoke along the same lines—were looking back on a century of extraordinary accomplishment, an epoch that had carried the physical sciences to a state of high development by the late years of the century.

The wavelike character of light had been demonstrated; the laws of electricity and magnetism were discovered and placed together in a unified framework; light was shown to be the manifestation of electric and magnetic field oscillations; the atomic hypothesis had increasingly taken hold as the century moved on; the laws of thermodynamics were successfully formulated and—for atomists—grounded in the dynamics of molecular motion; and more.

To be sure, although the gravitational and electromagnetic force laws seemed well understood, it remained yet to learn whether other kinds of forces come into play at the atomic level.

That is, there was work yet to be done, and not just at the sixth place of decimals. But a clocklike Newtonian framework seemed assured. In this *classical* picture of the physical world, space and time are absolute; and every bit of ponderable matter is at every instant at some definite place, moving with some definite velocity along some definite path, all governed by the relevant force laws according to Newton.

This classical outlook in fact continues to provide an excellent description of the physical world under conditions where velocities are small compared to the speed of light and relevant dimensions large compared to the size of atoms.

But our deeper conceptions of space-time have been transformed by relativity; and of objective reality, by quantum mechanics. Both run counter to everyday experience, to our common sense of the world. This is especially so for quantum mechanics, which is the focus of the present book.

Before we embark on our journey, it may be good in advance to sketch out very roughly some of the contrasts that will be encountered between the classical and quantum modes. For the most part here, we will be considering a system of point particles moving under the influence of interparticle and perhaps external force fields characterized by a potential energy function.

QUANTIZATION

Classically, a particle might be anywhere a priori; and it might have any momentum(momentum = mass × velocity).

Correspondingly, its angular momentum—a quantity defined in terms of position and momentum—might a priori have any value. So too the particle's energy, kinetic plus potential, might have any value above some minimum determined by the potential. Quantum mechanically, however, angular momentum can take on only certain discrete values.

It is *quantized*. Energy is sometimes quantized too, depending on details of the force field. It is this classically inexplicable discretization that provides the adjective "quantum" in quantum mechanics.

PROBABILITY

A much sharper and more profound contrast with classical mechanics has to do with the probabilistic character of quantum mechanics. For a classical system of particles, the state of affairs is completely specified at any instant by the position and momentum variables of all the particles.

The data on positions and momenta at any instant constitute what we may call the *state* of the system at that instant.

It tells all that can be known dynamically about the system. Other quantities of interest, such as energy, angular momentum, and so on, are defined in terms of the position and momentum variables.

Classical mechanics is deterministic in the sense that future states of the system are fully and uniquely determined if the state is specified at some initial instant.

The present determines the future. Of course, in practical situations the initial data will inevitably be compromised to some greater or lesser extent by measurement uncertainties.

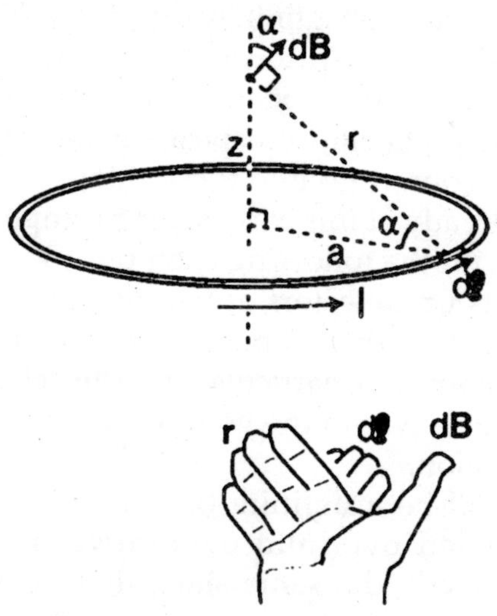

Fig. Out-of-the-Plane Field of a Circular Loop

Depending on the system under consideration, the future may or may not be sensitive to this uncertainty. But there is no limit *in principle* to the accuracy that can be imagined. There is no bar in principle, that is, to precise knowledge of the position and momentum of each particle, and therefore no bar to anticipating future developments. When wearing our classical, commonsense hats, we do not doubt that every bit of matter is at every instant at some definite place, moving with some definite momentum, whether or not we are there to observe.

The notion of state also arises in quantum mechanics. Here again the *state* of a system connotes *all that can possibly be known about the system at any instant.* Also, just as in the classical case, the system develops deterministically in that future states are fully determined if the state at some initial instant is given.

In this sense, here too the present determines the future. But there is a very profound difference. A quantum state does not precisely specify particle positions and momenta, it only specifies probabilities. Quantum mechanics, that is, is probabilistic!

For example, there are states for which the probability distribution of a particle's position is sharply localized, so that the position may be said to be almost definite(at the instant in question).

But there are other states for which the probability distribution is broad, so that upon measurement the particle might be found almost anywhere.

And there are infinitely many possibilities in between. So too for momentum: for some states the momentum is almost definite, for others it is broad, and there are infinitely many possibilities in between.

This probabilistic description obtains not because we have imperfect information about the state of the system, but is intrinsic. Moreover, the rules of probability composition have some very peculiar features. We will, of course, go into these things more fully later on, but it is important already at this early stage to emphasize a point that may be illustrated with the following example.

Suppose one sets up detectors at various locations to determine the position of a particle known(somehow) to be in a certain quantum state at a certain instant. If a particular detector clicks, we will have learned that the particle was in the volume occupied by that detector at the instant in question.

That is, there *will* be a definite finding of location. But if the experiment is repeated over and over, always with the particle arranged to be in exactly the same state, there will be a spread of outcomes. On different runs different detectors will click. Full knowledge of the quantum state does not allow one to predict the outcome event by event, only the probability distribution.

THE UNCERTAINTY PRINCIPLE

It is the case that any state that has a very localized probability distribution for position measurements will inevitably have a broad distribution for momentum measurements, and vice versa. There is a limit to how well one can jointly localize both position and momentum. So too for certain other pairs of *observables*(as measurable quantities are called). This is enshrined in the celebrated Heisenberg uncertainty principle. That principle is not some add-on to quantum mechanics; it is a technical consequence that flows from the structure of quantum mechanics. As must of course be the case, for the macroscopic objects of everyday life the Heisenberg limit is not at all a practical restriction.

We can, for example, know both the position and momentum of a moving jelly bean quite accurately enough for all everyday purposes. However, at the atomic level the uncertainty principle comes fully into play.

IDENTICAL PARTICLES

In the macroscopic world, we never encounter two or more objects that are strictly identical in every possible respect: mass, composition, shape, colour, electric charge, and so on. But even if we did—and we do at the microscopic level, where, for example, one electron is exactly the same as another—this would pose no conceptual problem for classical science. One can in principle keep separate track of the objects by, so to speak, pointing: object 1 is the one that's at this place, object 2 is the other one over there, and so on. For quantum mechanics this approach has its limits.

It is not possible to keep track in this way since locations are a probabilistic matter. Rather, there is a distinctly quantum mechanical approach to dealing with identity, one without classical analog. The implications are sometimes quite unintuitive, and they are profound. What is most remarkable is that all the known particles indeed come in strictly identical copies—all electrons are the same, all protons the same, and so on. Quantum field theory provides the only natural explanation for this striking fact of identity.

RADIOACTIVITY

This term refers to processes in which an atom *spontaneously* emits one or more particles: an alpha particle(helium nucleus) in the case of one class of processes, α decay; an electron(plus neutrino as we now know) in another class, β decay; an energetic photon in yet another class, γ decay. In α and β radioactivity, the parent atom is transmuted in the process into a daughter atom of a different chemical species.

There is no such transmutation in γ radioactivity. One speaks of any of these spontaneous events as a *decay* process. In the case of α

and β radioactivity there really is decay, the disappearance of the parent atom and its replacement by an atom of a different ilk. In radioactivity the atom does not change its chemical species membership; but as we will see later, it does undergo a change from one energy level to another. In that sense, here too there is decay—of the occupancy of the initial energy level. Not all atomic species are radioactive, but many are.

When radioactivity was first discovered around the end of the nineteenth century, there was great wonder and bafflement. Many questions were raised, among them the question: where in the atom(*if* in the atom) do the ejected particles come from?

This was clarified only after Rutherford formulated his famous model of the atom, picturing it as a swarm of electrons orbiting around a positively charged nucleus that is very tiny but that nevertheless carries most of the mass of the atom.

With that, it soon became clear that radioactivity is a *nuclear* phenomenon. Two other questions among many remained, and they were especially puzzling:(1) The emitted particles typically carry a lot of energy. Where does that energy come from?(2) How does the nucleus decide when to decay?

As to the first of these questions, the answer was already available in Einstein's 1905 formula $E = mc$; but it took a while before this sank in conceptually and before sufficiently accurate mass measurements of parent and daughter nuclei could be made to test the concept.

The deeper question(2) had to await the interpretative apparatus of quantum mechanics. If you take a collection of identical atoms of some radioactive species, you will find that the atoms do not all decay at some one characteristic instant but, rather, at various times—randomly. If the emissions are being detected by a counter, you may hear individual clicks as one or another atom decides to decay.

As time goes by there will of course be fewer and fewer surviving parent atoms.

As it turns out, the population of survivors decreases with time in an essentially exponential fashion, the average time(or, briefly, the *lifetime*) being characteristic of the particular species under consideration. On the classical outlook, the problem is this. The atoms of the given species are presumed to be identical. If they are governed by the clockwork regularity of classical science, why don't they all decay at the same instant, whatever may be the mechanism that causes radioactive decay?

The quantum mechanical answer is that the world is a probabilistic place. An ensemble of identical atoms starting in identical conditions will distribute their decays in a probabilistic way over time. One cannot predict what will happen event by event, atom by atom.

What *can* be deduced quite generally is the exponential character of the decay curve. But the mean lifetime varies from species to species and depends sensitively on details of the underlying quantum dynamics. It should be said here that the traditional classes of nuclear instability, α, β, and γ, are only three among a much wider range of decay processes that occur in nature, including hordes of reactions involving subnuclear particles: pi meson decay, muon decay, and so on.

The average lifetimes vary over an enormous range, from roughly $10^{''}$ seconds for certain subnuclear particles to billions of years and more for certain α emitters(among these, U, whose half-life happens to be about the same as the age of the earth).

TUNNELING

The probabilistic structure of quantum mechanics incorporates the possibility that a particle can be found in locations that are absolutely forbidden to it classically. For example, it can happen classically that there is an energy barrier that separates one region of space from another, so that particles below some energy threshold cannot penetrate the barrier and thus cannot move from one region to the other(it may take more energy than you've got to climb the hill that intervenes between where you are and where you want to go).

Quantum mechanically, there is a finite probability that such strange things can happen. Particles can be found in, and can *tunnel* through, classically forbidden regions.

ANTIMATTER

In attempting to find a relativistic generalization of Schroedinger's quantum equation for the electron, P. A. M. Dirac devised a theory that was spectacularly successful in its application to the hydrogen atom but that carried with it some seemingly bizarre baggage: among other things, negative energy states for the free electron.

When properly reinterpreted this transformed itself into the prediction of a new particle having the same mass as the electron but opposite(that is, positive) charge. The antielectron, or *positron* as one calls it, was soon discovered experimentally.

The situation has since become generalized. Relativistic quantum theory predicts that particles having electric charge must come in pairs with opposite charges but identical masses(and identical lifetimes if unstable).

One member of the pair is called the particle, the other the antiparticle. Which is called by which name is a matter of history and convenience. It turns out that there are other kinds of "charge" in addition to electric charge; for example, so-called baryon number charge.

The necessity of particle-antiparticle pairs obtains for charges of any kind. Thus, not only is there an antiproton to the proton, there is an antineutron to the neutron. The neutron is electrically neutral but it has baryon number charge. On the other hand, the photon and π meson among others do not have antiparticles; or as one says, each is its own antiparticle.

CREATIONISM, DESTRUCTIONISM

Our notion of what it means to say that something is made of other things has undergone a revolutionary transformation in this century. When you take a clock apart you find gears, springs, levers, and so on(or maybe a quartz crystal and battery).

You say the clock is made of these parts. If you take apart the parts in finer and finer detail, you eventually get to atoms. If you take apart atoms, there are electrons and nuclei of various sorts.

Going on, you find that the nuclei are made of protons and neutrons, and then that these are made of quarks and gluons. At the microscopic level, incidentally, taking apart means zapping the target with a projectile and looking at the pieces that emerge. In earlier years the surprise may have been that deconstruction did not stop at the atom. Still, the ancient notion could persist that, eventually, one comes to the immutable ingredients of the world, building blocks that can arrange and rearrange themselves in various combinations but that are themselves eternal and indestructible.

Thus, for example, the nuclear reaction $d + t \rightarrow \mathrm{He} + \mathrm{n}$ can be pictured as a mere rearrangement of the neutron(n) and proton(p) ingredients of the deuterium(d) and tritium(t) nuclei, the ingredients reemerging as the helium nucleus(He) with one neutron left over.

The particle reaction(i) $\pi + p \rightarrow 3 + K$ might be taken to indicate that the particles involved here— pion, proton, lambda particle, kaon— are made of tinier things, perhaps quarks, that are similarly rearranging themselves.

But if so, what does one make of the reaction(ii) $\pi\ Cp \rightarrow 3\ CKC\pi$, in which an extra pion appears on the right? Haven't the quarks already been conceptually "used up" to account for reaction(i), so that there are no ingredients left over to explain reaction(ii)?

And what does one make of the reaction $p + p \rightarrow p + p + \pi$? No amount of rearrangement can explain how it is that the final system contains the same objects as the initial system *plus* something else. There is no getting around it, the π is simply created here de novo; or at any rate its ingredients are. In short, down at the subnuclear level one is simply forced to acknowledge that particles can be created and destroyed!

This creation and destruction of matter is not something of everyday experience. It is a phenomenon that comes into play at high-energy particle accelerators, in the collisions induced by cosmic rays(high-energy particles that rain on the earth from outer space), in the stars and wider cosmos, and in certain radioactive decay processes. The transactions underlying most of science, technology, and everyday life have mostly to do with the "mere" motions and rearrangements of electrons and nuclei.

However, there is one very notable exception to this, even in everyday life. It involves a thoroughly familiar phenomenon interpreted in a modern light, namely, light! A beam of light is nothing but an assemblage of massless particles, *photons,* moving at(what else?) the speed of light. Because they are massless, photons are easy to create. They are created whenever the light switch is turned on.

Regarded microscopically, what happens is that they are produced in electron and atomic collision processes taking place in the light source when the latter is heated or otherwise "excited." Photons are destroyed when they impinge on and are absorbed by nontranslucent material bodies(walls, books, the retina of the eye, etc.).

Photon creationism-destructionism actually entered the world when Einstein proposed his particle-like interpretation of electromagnetic radiation. But the photon concept had a protracted birth, and the photon is anyhow such a special particle.

It is massless; it is the quantum of a field we have known classically. Somehow, for photons, the enormity of creation-destruction as such did not seem to attract much philosophical discussion in the early years of this century.

In any case, for a while one could still cling to the idea that "real" *ponderable* particles, particles with nonzero mass such as electrons, protons, and neutrons, are truly immutable. But there is no such immutability for them either.

This first became apparent with the discovery of the neutron and the recognition of its role in nuclear beta decay. The neutron is destroyed, the proton, electron, and antineutrino created. The antineutrino, which is highly unreactive, easily escapes the nucleus and passes through the earth, the solar system, the galaxy, and into outer space without leaving much of a scratch. But that's another story. Where does quantum theory fit in? The quantum theory of the electromagnetic field got its start in the heroic period of the mid 1920s when the foundations of quantum mechanics were being established. Quantum electrodynamic theory was designed from the beginning to account for photon creation and destruction.

The photon emerges naturally in the theory as a quantum of the electromagnetic field. Since that time physicists have brazenly invented other fields, fields not known to us in their classical guise but that are invented for the purpose of being quantized to yield other particles as well. So, for example, there is a field that makes and destroys electrons. The older theories used to have separate fields as well for protons, neutrons, pions, and so on. We have now reached a more basic level involving, among other entities, quarks and gluons. But these too can be created and destroyed.

Beginnings

In its modern form, the structure of quantum theory was laid down in the middle of the 1920s in a concentrated burst of creativity and transformation that is perhaps without parallel in the history of scientific thought.

Mainly, the creators were very young: Werner Heisenberg, Paul Dirac, Pascual Jordan, and Wolfgang Pauli were all in their twenties. The elders included Erwin Schroedinger, who published his famous wave equation at age thirty-nine, and Max Born, who at the age of forty-three recognized and helped elaborate what Heisenberg had wrought.

The new outlook brought with it an unintuitive concept of reality along with a number of attendant oddities of various sorts. Among contemporary physicists, some could not readily absorb the new doctrine. They grumbled and fell out. But already the earliest applications to phenomena met with convincing success. Informed dissidents, Albert Einstein foremost among them, soon accepted the effective correctness of quantum mechanics. They were reduced to hoping that classical reality prevails at some deeper level of nature not readily accessible to observation. That deeper level, if there is one, is still today nowhere in sight.

As far as the eye can see, the principles of quantum mechanics stand irreducible and empirically unchallenged.

In cases where the difficult experiments and corresponding theoretical calculations can be carried out with high precision, quantitative agreement is spectacular.

As often happens in intellectual revolutions, it was the younger generation that could adapt to the new ways of thinking somewhat more easily than the older one. Succeeding generations have had an even easier time of it; they simply grew up with the subject.

Nevertheless, the world view of quantum mechanics is odd; and the oddest thing of all is that, still today, many decades after its foundation, quantum mechanics continues to seem odd even to scientific practitioners who work with the subject every day and who know and operate confidently in its framework. Their wonderment expresses itself not so much at the operational level as at a philosophical one. Deep questions persist at that level. We will surely not resolve them here.

The more modest aim here is simply to convey some notion of what quantum mechanics is: its principles and some of its consequences and oddities.

Many questions within the classical framework were still unresolved toward the end of the nineteenth century, especially questions having to do with the nature of atoms—and for some diehards, even the very existence of atoms.

But the Newtonian framework was not in doubt. It is possible today in hindsight to recognize hints of quantum effects, empirical departures from classical expectation that should have been pounced on by our nineteenth century ancestors. However, this is only in hindsight. They *did* in fact encounter anomalies and *did* fret over them, but it was far from clear at the time that these could not be resolved within the still developing classical picture.

There are vast stretches of contemporary macroscopic science and engineering that still do very well today without any reference at all to the quantum mechanical basis of nature. This is so because classical Newtonian behaviour emerges for the most part as a very good approximation to quantum mechanics for macroscopic systems. But this assertion has to be understood in a qualified sense.

The qualification can be illustrated by means of an example. Consider the flow of oil through a smooth cylindrical pipe, the flow being driven by a pressure differential that is established between the ends of the pipe. If the pressure differential is not too large the flow will be smooth; and it is then an easy matter, a standard textbook problem in classical fluid dynamics, to compute the flow rate, the volume of oil transported per unit time. The answer depends on the length and diameter of the cylinder and on the pressure differential. These are parameters of experimental choice or circumstance. But the answer also depends on the viscosity of the oil.

If the value of that parameter is simply accepted as a given fact of nature, as a quantity to be determined empirically, then the computation of flow rate may be said to proceed along purely classical lines without reference to quantum mechanics.

However, to understand why oil has the viscosity and other properties that it has, one has to move down to the atomic level. And there the differences between quantum and classical science are as striking as can be. Another qualification should be noted.

The quantum mechanical rules, the concrete equations, are definite and well established. *In principle* one can compute the structure of oil molecules and work out the way these molecules interact among themselves in bulk oil and thence go on to the viscosity of oil. But a completely detailed calculation that traverses the whole route from the individual molecule and its ingredients all the way up to the astronomical number(about 10)of molecules present in even a small drop of oil is utterly unthinkable.

The single molecule is already complicated enough. Thus, approxiinations and aggregate treatments have to be adopted along the way, relying on various rich and active fields of scientific inquiry; for example, the field of statistical mechanics.

A pumper who wants highly accurate predictions of flow rate is well advised to adopt the empirical value of viscosity. But that same pumper may also share with others a curiosity about why things are the way they are.

Moreover, there is the possibility of learning enough at the microscopic level to design molecular additives that can alter the viscosity in wanted directions. As with viscosity, so too for other kinds of information that enter in parametric form into the various branches of classical science and engineering: tensile strength of materials, thermal conductivity, electrical resistance, equations of state(the relation of pressure to density and temperature) for various gases and liquids, optical reflection coefficients, and so on.

The different fields have their independent methodologies and concepts. None suffers any shortage of engaging intellectual and practical challenges within its own framework. But so far as we know, science is seamless.

At a deeper level the different fields share in common the science of atoms, where the quantum reigns. Deeper still is the fantastic world of the subatomic particles; and farther out, the world of the cosmos. Quantum mechanics first began to intrude itself on mankind's attention in the very first year of the twentieth century. It did not by any means spring up full grown.

The beginnings can be sharply placed within a rather esoteric corner of the scientific scene of those times; namely, the physics of *blackbody radiation*. The blackbody question has to do with the frequency spectrum of electromagnetic radiation that fills any volume of space surrounded by material walls in thermal equilibrium. That seems an awfully specialized topic.

However, it had been established decades earlier through elegant thermodynamic reasoning that the spectrum, the radiation intensity as a function of frequency, must be of a fundamental character. It can depend only on frequency and temperature, not on the shape of the vessel nor, more strikingly, on the kinds of materials that the walls are made of. Deep issues therefore appeared to be at stake.

Experimental measurements over various parts of the frequency spectrum were actively pursued toward the end of the century. The challenge on the theoretical side was to predict the spectrum. It was the German physicist Max Planck who succeeded.

That was in the fall of 1900. We will describe the scientific issues more fully later on; but briefly, what happened was this. Presented with the latest experimental results on the blackbody spectrum, Planck sat down at one point and in not much more than an evening's work so far as we know, he devised—stumbled upon—an empirical formula that fit the spectral data remarkably well.

This was something more than a case of raw curve fitting, however, since he brought to the task some guiding ideas that had emerged from earlier work by himself and others. Nevertheless, his formula was essentially empirical.

Over the succeeding months he sought to deduce it within the framework of the classical theory of his times. This required some statistical mechanics reasoning.

But the statistical mechanics aspects of classical science were still somewhat in flux and Planck did not recognize, or at any rate did not choose to follow, a simple path to the blackbody spectrum that was available to him.

Had he taken that path(noticed slightly earlier by Lord Rayleigh), he would have encountered catastrophic disagreement with the data. Instead, he followed a more complicated route that was mostly classical in its outlines, but then did some fiddling that we will describe later on. Out came the empirical Planck blackbody formula! From this small seed came the quantum revolution.

There was no immediate commotion in the streets. Only a small band of scientists were participating in or paying close attention to these developments. Among those few it was pretty clear that something new was afoot, but it was far from clear what that new thing was. A decisive insight was provided by Albert Einstein in 1905, the miracle year in which, among other things, he published his papers inaugurating the special theory of relativity.

What Einstein drew from Planck's discovery was the startling hypothesis that electromagnetic radiation of frequency f can exist only in discrete energy bundles, *quanta,* and that the energy of each such bundle is proportional to the frequency: energy = hf, where the proportionality constant h is the new parameter of nature that had entered into Planck's blackbody formula.

These quanta of Einstein are particle-like entities that have since come to be called *photons*. However, light is nothing but a form of electromagnetic radiation; and one of the triumphs of nineteenth century science had been the discovery that light is a wavelike phenomenon. Here then, with Einstein's quanta, was the beginning of the celebrated wave particle duality conundrum that hovered over physics during the next two decades.

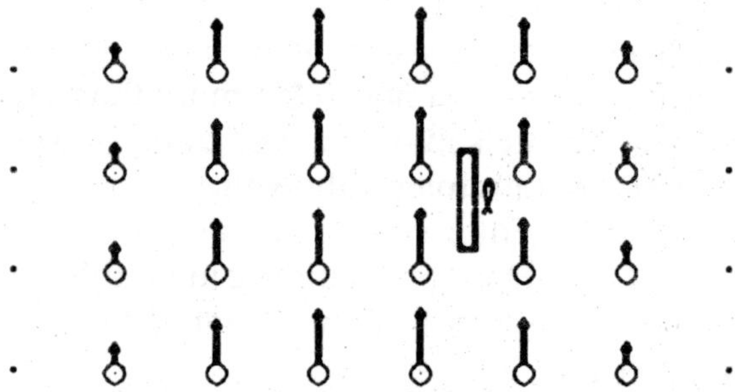

Fig. A Classical Calculation of the Momentum of a Light Wave. An Antenna of Length is Bathed in an Electromagnetic Wave. The Black Arrows Represent the Electric Field, the White Circles the Magnetic Field Coming out of the Page. The Wave is Traveling to the Rright.

Quantum ideas were soon extended from radiation to ponderable matter. In fact, Planck's work had already suggested some sort of energy quantization for ponderable matter; but, excusably for that pioneering effort, the suggestion was rather murky.

Following up on these hints, in 1907 Einstein developed a simple quantum model of the specific heat of material bodies. Specific heat is a parameter that characterizes the temperature change induced in a material body when it absorbs a given quantity of heat energy. Einstein proceeded as follows. Material bodies can of course sustain sound waves over some range of frequencies f.

For these he adopted the same quantization hypothesis that he had adopted for electromagnetic radiation; namely, the assumption that the energy in a sound wave disturbance of frequency f can come only in bundles of energy hf. He was content to take a single representative frequency. Others soon generalized to cover the whole frequency range. The model provided a qualitatively successful account of certain anomalies, departures from the expectation of classical theory, that had been known empirically for some time. The band of scientists paying attention to quantum developments began to grow.

In 1913 the young Danish physicist Niels Bohr turned to the inner workings of the atom. What might the developing quantum ideas have to say on this subject?

For the content and structure of the atom he took up a model that had been convincingly proposed only a couple of years earlier by the great experimentalist Ernest Rutherford. In it the atom is pictured as a kind of miniature solar system: a tiny, positively charged nucleus at the centre(analog of the sun), and very much lighter, negatively charged electrons(the planets) orbiting around the nucleus.

Rutherford came to this picture of the atom through a celebrated experiment in which his colleagues H. Geiger and E. Marsden bombarded a thin metal foil with fast alpha particles and observed, to their wonderment and Rutherford's, that the alpha particles occasionally scattered through large angles.

Collisions with the atomic electrons, which are very tiny in mass, could not be expected to produce substantial deflections of the fast, heavier alpha particles. But a heavy, highly concentrated positive charge, an atomic nucleus, would do the trick. On this picture, Rutherford could work out the expected distribution of scattering angles, proceeding along classical Newtonian lines based on the Coulomb law of force between charged particles.

The result agreed well with experiment and confirmed Rutherford in his model of the atom. But the Rutherford atom presented a conundrum. To illustrate, consider the simplest atom, hydrogen.

It has a single electron moving around a proton nucleus. The electron, acted on by the Coulomb force of the nucleus, is in a state of accelerated motion. According to the classical laws of electricity and magnetism, an accelerating charge must constantly be emitting electromagnetic radiation and thereby losing energy. Suppose for a moment that this energy loss can be ignored. Then, classically, the electron travels in an elliptical orbit with a revolution frequency that depends on the electron energy among other things.

It radiates at the frequency of that orbital motion. But there are infinitely many *possible* orbits, just as in the case of objects(planets, comets, asteroids, spaceships) moving around the sun. Given a macroscopic collection of hydrogen atoms, it would be surprising if the electrons in the different atoms were not traveling in a whole range of different orbits. That is, on this picture one would expect an essentially continuous spread of radiation frequencies.

In fact, however, atoms radiate only at certain discrete frequencies, in a characteristic pattern that distinguishes one species of atom from another(one speaks of the characteristic frequencies as "lines" since they show up as lines in a spectrographic display). An even more serious problem for the classical Rutherford atom is that one is not really allowed to ignore the fact that the electron is losing energy as it radiates.

Instead of traveling steadily on an elliptical orbit, therefore, a classical electron must eventually spiral into the nucleus, its orbital frequency and thus the radiated frequency changing all the while as the orbit shrinks in size. Empirically, however, nothing like this makes either spectroscopic or chemical or common sense. Confirmed atomists had in fact been confronted with these paradoxes for a long time, trying to figure out how it is possible, classically, to stabilize atoms against radiative collapse; also, how to account for their discrete line spectra. Here, presented in a series of steps, is what Bohr did to resolve the conundrum, at least for the one-electron atom.

Step 1: Ignore radiation for the moment and work out the electron orbits using purely classical dynamics, as discussed above. Bohr restricted himself to circular orbits.

Step 2: Now impose a "quantum condition" devised by Bohr to determine which orbits are quantum mechanically "allowed," all others simply being forbidden! A consequence of this will be that only certain energies are possible. Instead of spanning a continuous range of possible values the allowed energies now form a discrete set; they are *quantized*.

Step 3: Assert that the electron does not radiate while moving in one of these allowed orbits. But when the electron happens to be in an excited level of energy E and "decides" to jump to a lower level of energy $E0$, it emits a photon of frequency f determined by the equation $hf = E - E0$. This equation is arranged to insure energy conservation, since according to Einstein hf is the photon energy.

Bohr invented his rules very soon after learning of a remarkably simple empirical formula that the Swiss schoolteacher, Johann Jakob Balmer, had devised many years earlier for the frequencies of the hydrogen atom. Balmer's formula, which involved only a single adjustable parameter(the "Rydberg"), predicted that there should be infinitely many hydrogen lines. Only several of the lines were known in Balmer's time, many more when Bohr turned to the subject.

There can be no doubt that Bohr tailored his quantum rules to fit the facts. But the remarkable thing is that he *could* fit the facts, that his simple but classically inexplicable rules worked. Bohr could determine the Rydberg solely in terms of basic parameters that were already known and over which he had no freedom to make adjustments; namely, the charge and mass of the electron, and Planck's constant h. The agreement with experiment was very good indeed.

A vigorous and greatly broadened era of quantum theory now got under way as physicists sought to expand Bohr's beachhead to cover the effects of external electric and magnetic fields on the energy levels of hydrogen, to incorporate relativistic effects, to apply quantum ideas to multielectron atoms, and so on.

Bohr's quantum conditions were speculatively generalized to cover this wider range of questions. Just as in Bohr's original formulation, the generalized rules had an ad hoc character: quantum conditions superimposed on top of classical reasoning without any deeper understanding of where those quantum conditions come from. To a considerable extent, developments were guided by the so-called *correspondence principle,* which had been formulated and exploited by Bohr and then taken up by others. Roughly, it is the notion that quantum behaviour must resemble classical behaviour for large energies. This idea was adopted and then ingeniously(and nervily) pressed into service for all energies.

There were failures, but there were also many successes. It was a zany era of progress and confusion, a hodgepodge of inexplicable quantum rules and classical dynamics. It flourished for about a dozen years, the interval between Bohr's 1913 papers and the birth of modern quantum theory. The physicist Isidor Rabi, looking back, described it as a time of "artistry and effrontery." The modern theory began along two seemingly unrelated lines, one opened up by Heisenberg, the other independently by Schroedinger.

The pace was breathtaking. The first steps were taken by Heisenberg on a holiday in the spring of 1925. Although constrained and indeed guided to some extent by the correspondence principle, he broke sharply with the concepts of classical mechanics at the atomic level. He argued for abandoning the notion of definite positions and momenta on the ground that these are basically unobservable at that microscopic level. But atomic energy levels *are* observable through their role in determining the frequencies of atomic lines.

Heisenberg set up a new mechanics aimed at that target. What he postulated seemed to come out of the blue; and it was expressed in a mathematical language that was unfamiliar to many, even to Heisenberg himself. However, it had the air of being on the right track. Heisenberg's mentor at Göttingen, Max Born, received the paper favorably, puzzled a while over the mathematics, then recognized it for what it was. Within a few brief months, by September, he and another assistant, Fascual Jordan, completed a paper extending Heisenberg's ideas and identifying his mathematical objects as *matrices*.

The story is told—if true, it says something about the times—how the then unknown Jordan came to work with Born. The young man found himself traveling in a railroad compartment with Born and a colleague of Born's. Born was talking to his colleague about matrices. Jordan overheard, introduced himself, and said that he knew about matrices and maybe could help. Born signed him on, just like that! Their joint paper was produced not much later.

Soon after, in November, Heisenberg joined Born and Jordan to produce the celebrated "three-man" paper(*Dreimanner Arbeit*) which set out in an extended and logical framework Heisenberg's quantum theory, now dubbed *matrix mechanics*. Meanwhile, basing himself only on Heisenberg's original paper and unaware of the work of Born and Jordan, Paul Dirac in Cambridge similarly extended Heisenberg's ideas, in a different, elegant mathematical language.

It brought out the formal similarities between quantum and classical mechanics, and also the differences. Before the year was out Pauli had already applied the new quantum theory to the hydrogen atom. In particular, he successfully worked out the effect of an electric field on the energy levels of hydrogen, a problem that could not be tackled in the old quantum theory.

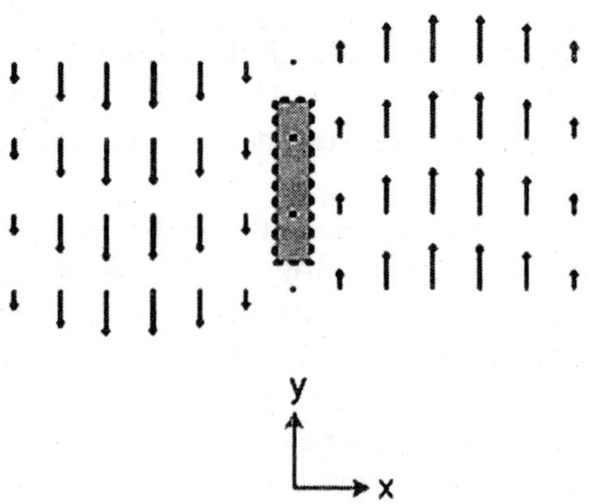

Fig. The Magnetic Field of the Wave. The Electric Field, not Shown, is Perpendicular to the Page.

All of this was in the space of not much more than half a year! And then, in the very first month of the next year, 1926, there came the first of Schroedinger's papers laying out what looked to be an entirely different quantum theory.

Schroedinger built on an idea that had been introduced several years earlier in the doctoral dissertation of Louis de Broglie, who was by then almost an elder at age thirty!

What de Broglie suggested was that just as light had been shown to be both wavelike and particle-like, so too perhaps there are "matter waves" somehow associated with ponderable matter, for example, electrons. Einstein recognized the promise in this idea and gave it his influential blessing. Schroedinger extended it into a fullblown theory. Pursuing analogies with classical mechanics and optics, he introduced the idea of a wave function that is to be associated with any system of material particles; and he wrote down an equation that the wave function must satisfy, all of this even though the physical meaning of this function was initially quite vague.

No matter that it was vague, however. The equation passed a first and by now mandatory test. It produced the right energy levels for the nonrelativistic hydrogen atom. Except for some initial reserve, even grumpiness, on the part of Heisenberg and others at Göttingen, Schroedinger's papers quickly captivated the world of physics. Unlike matrix mechanics, his wave mechanics was expressed in a familiar mathematical language; and, initially, it had about it the air of a theory that might be reconciled with classical notions of reality. That latter proved to be an illusion.

If a vote had been taken at the time to choose between the two theories it is probable that most physicists would have boycotted the election altogether(a pox on both of these newfangled quantum theories!). Among the voters, however, it is likely that the majority would have opted for wave over matrix mechanics.

But it soon transpired that these two theories are really one and the same, as Schroedinger could demonstrate convincingly enough and as others could soon prove to higher standards of mathematical rigor. The two theories, that is, are just two different mathematical *representations* among an infinite number of other, possible representations of the same physics.

This is not altogether unlike the case of different coordinate systems being used to describe the same phenomena but from different vantage points. The principles of quantum theory can in fact be formulated in highly abstract terms that do not commit to any particular representation. However, both for practical calculations and for purposes of developing an intuitive feel for quantum mechanics, it is usually best to come down from the abstract heights. It will be most convenient in the present exposition to proceed along the Schroedinger line.

Quantum mechanics was taken up widely and quickly following the papers of the founders. The earliest applications concentrated on various energy level problems. It was possible to address this class of problems without facing up to broader interpretative questions; in particular, questions having to do with the physical significance of the Schroedinger wave function. The modern interpretation was supplied soon enough, however, beginning with a remark made by Born in a 1926 paper on the quantum theory of scattering. This was swiftly elaborated.

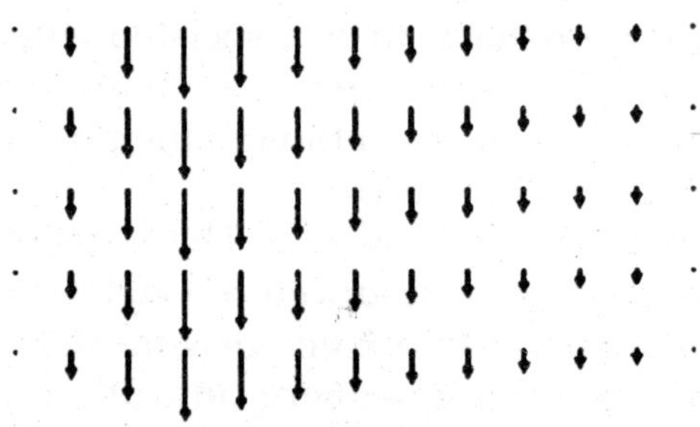

Fig. A Nonsinusoidal Wave.

Above all others, it was Niels Bohr who presided over development of the general interpretative principles of quantum mechanics. What emerged was the picture of a probabilistic structure of nature and hence a sharp break with intuitive notions of reality.

Among the giants, Schroedinger himself resisted, as did Einstein. Einstein watched with "admiration and suspicion." For a time he pressed his antiprobabilistic outlook("God does not play dice") in a celebrated series of debates with Bohr. Bohr won. Einstein eventually accepted the correctness of quantum mechanics as far as it goes; but for the rest of his life he held out for the existence of a deeper, not yet accessible, level of classical reality.

What does the wave function signify? Everything. According to the principles of quantum mechanics the wave function incorporates all that can be known about the state of the system at any instant. But it does not in general tell where the particles are located or what their momenta are.

What it gives us, and that's all we can know, are *probabilities* concerning the outcomes of various kinds of measurements that might be made on the system, measurements of position, momentum, energy, angular momentum, and so on. The contrast with classical language is interesting here. For example, a classical scientist will write "let x denote the position of the particle," rather than "let x denote the outcome of a *measurement* of the position of the particle."

Classically, unless one is concerned with the practicalities of a measurement, it will be understood that the particle surely *is* somewhere. Yes, its position variable *can* in principle be measured, but there is no need to emphasize the latter point or speak of measurement.

Quantum mechanically, on the other hand, the particle is *not* at some definite place, not unless a measurement reveals it to be at that place. One can speak only of probabilities in connection with a measurement of position and other variables. The notion of measurement, therefore, is nearer to the surface in quantum mechanics.

Heisenberg: "We can no longer speak of the behaviour of the particle independently of observation." Bohr: "An independent reality can neither be ascribed to the phenomena or the agencies of observation." Three baseball umpires: First umpire, "I calls them the way I sees them." Second umpire, "I calls them the way they *are*."

Third umpire, "They ain't nothing till I calls them." Let us return briefly to the historical story. Schroedinger's version of quantum mechanics brought out clearly the waveparticle duality aspect of ponderable matter. Wave-particle duality for electromagnetic radiation, whose particle-like aspect is the photon, found its proper quantum basis in 1927 with the application of quantum principles to the electromagnetic field.

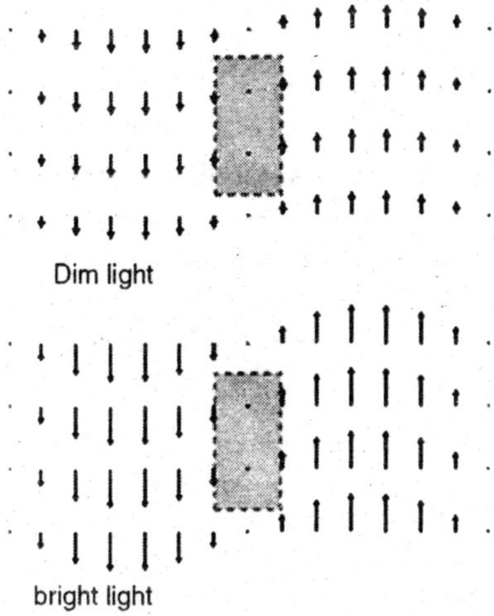

Fig. Bright and Dim Light Travel at the Same Speed.

This was the work of Paul Dirac, who inaugurated *quantum electrodynamics* in a paper published that year. Dirac struck again in the following year, 1928, with his relativistic wave equation of the electron. Apart from an unsuccessful early attempt to marry his quantum ideas to special relativity, Schroedinger's quantum theory had addressed itself to nonrelativistic situations, situations where velocities are small compared to the speed of light.

Dirac succeeded in constructing a relativistic quantum theory of the electron, a theory that incidentally(!) predicted the existence of antiparticles—although Dirac did not initially recognize that implication.

Chapter 2

Magnetism and its Properties

In physics, magnetism is one of the phenomena by which materials exert attractive or repulsive forces on other materials. Some well-known materials that exhibit easily detectable magnetic properties(called magnets) are nickel, iron, cobalt, and their alloys; however, all materials are influenced to greater or lesser degree by the presence of a magnetic field.

Magnetism also has other definitions/descriptions in physics, particularly as one of the two components of electromagnetic waves such as light. Aristotle attributes the first of what could be called a scientific discussion on magnetism to Thales, who lived from about 625 BC to about 545 BC.

Around the same time in ancient India, the Indian surgeon, Sushruta, was the first to make use of the magnet for surgical purposes.

In ancient China, the earliest literary reference to magnetism lies in a 4th century BC book called *Book of the Devil Valley Master*: "The lodestone makes iron come or it attracts it."

The earliest mention of the attraction of a needle appears in a work composed between AD 20 and 100(*Louen-heng*): "A lodestone attracts a needle." The ancient Chinese scientist Shen Kuo(1031-1095) was the first person to write of the magnetic needle compass and that it improved the accuracy of navigation by employing the astronomical concept of true north(*Dream Pool Essays*, AD 1088), and by the 12th century the Chinese were known to use the lodestone compass for navigation.

Alexander Neckham, by 1187, was the first in Europe to describe the compass and its use for navigation. In 1269, Peter Peregrinus de Maricourt wrote the *Epistola de magnete*, the first extant treatise describing the properties of magnets.

In 1282, the properties of magnets and the dry compass were discussed by Al-Ashraf, a Yemeni physicist, astronomer and geographer. In 1600, William Gilbert published his *De Magnete, Magneticisque Corporibus, et de Magno Magnete Tellure*(*On the Magnet and*

Magnetic Bodies, and on the Great Magnet the Earth). In this work he describes many of his experiments with his model earth called the terrella. From his experiments, he concluded that the Earth was itself magnetic and that this was the reason compasses pointed north(previously, some believed that it was the pole star(Polaris) or a large magnetic island on the North Pole that attracted the compass).

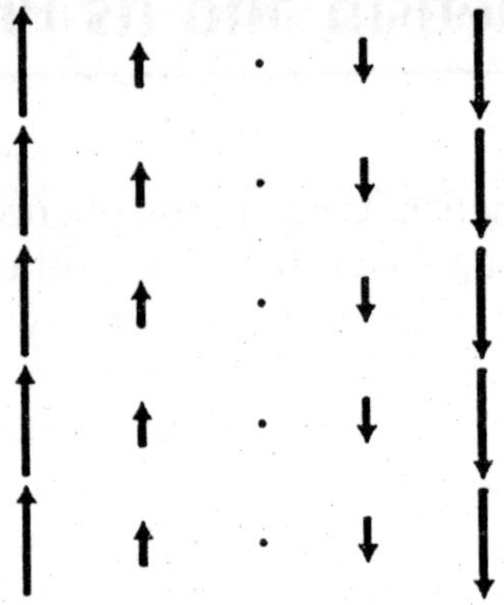

Fig. Multiplying the Field by -1

An understanding of the relationship between electricity and magnetism began in 1819 with work by Hans Christian Oersted, a professor at the University of Copenhagen, who discovered more or less by accident that an electric current could influence a compass needle.

This landmark experiment is known as Oersted's Experiment. Several other experiments followed, with André-Marie Ampère, Carl Friedrich Gauss, Michael Faraday, and others finding further links between magnetism and electricity.

James Clerk Maxwell synthesized and expanded these insights into Maxwell's equations, unifying electricity, magnetism, and optics into the field of electromagnetism. In 1905, Einstein used these laws in motivating his theory of special relativity, requiring that the laws held true in all inertial reference frames.

Electromagnetism has continued to develop into the twentieth century, being incorporated into the more fundamental theories of gauge theory, quantum electrodynamics, electroweak theory, and finally the standard model.

MAGNETS AND MAGNETIC MATERIALS

A magnet is a material or object that produces a magnetic field. This magnetic field is invisible and causes the most notable property of a magnet: a force that pulls on nearby magnetic materials, or attracts or repels other magnets. The structure of the invisible magnetic field of a magnet is made visible by the pattern formed when iron filings are scattered around the magnet, as in the accompanying figure. A "hard" or "permanent" magnet is one that stays magnetized, such as a magnet used to hold notes on a refrigerator door.

Permanent magnets occur naturally in some rocks, particularly lodestone, but are now more commonly manufactured. A "soft" or "impermanent" magnet is one that loses its memory of previous magnetizations.

"Soft" magnetic materials are often used in electromagnets to enhance(often hundreds or thousands of times) the magnetic field of a wire that carries an electric current and is wrapped around the magnet; the field of the "soft" magnet increases with the current.

Two measures of a material's magnetic properties are its magnetic moment and its magnetization. A material without a permanent magnetic moment can, in the presence of magnetic fields, be attracted (paramagnetic), or repelled(diamagnetic). Liquid oxygen is paramagnetic; graphite is diamagnetic.

Paramagnets tend to intensify the magnetic field in their vicinity, whereas diamagnets tend to weaken it. "Soft" magnets, which are strongly attracted to magnetic fields, can be thought of as strongly paramagnetic; superconductors, which are strongly repelled by magnetic fields, can be thought of as strongly diamagnetic.

MAGNETIC FIELD

The magnetic field(usually denoted B) is called a field(physics) because it has a value at every point in space. The magnetic field(at a given point) is specified by two properties:(1) its *direction,* which is along the orientation of a compass needle; and(2) its *magnitude*(also called *strength*), which is proportional to how strongly the compass needle orients along that direction.

Direction and magnitude makes B a vector, so B is a vector field.(B can also depend on time.) In SI units the strength of the magnetic field is given in teslas.

MAGNETIC MOMENT

A magnet's magnetic moment(also called magnetic dipole moment, and usually denoted μ) is a vector that characterizes the magnet's overall magnetic properties.

For a bar magnet, the direction of the magnetic moment points from the magnet's north pole to its south pole, and the magnitude relates to how strong and how far apart these poles are. In SI units the magnetic moment is specified in terms of A·m^2. A magnet both produces its own magnetic field and it responds to magnetic fields. The strength of the magnetic field it produces is at any given point proportional to the magnitude of its magnetic moment.

In addition, when the magnet is put into an "external" magnetic field produced by a different source, it is subject to a torque tending to orient the magnetic moment parallel to the field.

The amount of this torque is proportional both to the magnetic moment and the "external" field. A magnet may also be subject to a force driving it in one direction or another, according to the positions and orientations of the magnet and source.

If the field is uniform in space the magnet is subject to no net force, although it is subject to a torque. A wire in the shape of a circle with area A and carrying current I is a magnet, with a magnetic moment of magnitude equal to IA.

MAGNETIZATION

The magnetization of an object is the local value of its magnetic moment per unit volume, usually denoted M, with units A/m. It is a vector *field*, rather than just a vector(like the magnetic moment), because the different sections of a bar magnet generally are magnetized with different directions and strengths.

A good bar magnet may have a magnetic moment of magnitude 0.1 A·m^2 and a volume of 1 cm^3, or 0.000001 m^3, and therefore an average magnetization magnitude is 100,000 A/m. Iron can have a magnetization of around a million A/m. Such a large value explains why magnets are so effective at producing magnetic fields.

THE TWO MODELS FOR MAGNETS

MAGNETIC POLE MODEL

Although for many purposes it is convenient to think of a magnet as having distinct north and south magnetic poles, the concept of poles should not be taken literally: it is merely a way of referring to the two different ends of a magnet.

The magnet does not have distinct "north" or "south" particles on opposing sides.(No magnetic monopole has yet been observed.) If a bar magnet is broken in half, in an attempt to separate the north and south poles, the result will be two bar magnets, *each* of which has both a north and south pole.

The magnetic pole approach is used by most professional magneticians, from those who design magnetic memory to those who design large-scale magnets. If the magnetic pole distribution is known, then outside the magnet the pole model gives the magnetic field exactly. By simply supplementing the pole model field with a term proportional to the magnetization the magnetic field within the magnet is given exactly. This pole model is also called the "Gilbert Model" of a magnetic dipole.

AMPÈRE MODEL

Another model is the "Ampère Model", where all magnetization is due to the macroscopic effect of microscopic, or atomic, "bound currents", also called "Ampèrian currents". For a uniformly magnetized bar magnet in the shape of a cylinder, with poles uniformly distributed on its ends, the net effect of the microscopic bound currents is to make the magnet behave as if there is a macroscopic sheet of current around the cylinder, with local flow direction normal to the cylinder axis.(Since scraping off the outer layer of a magnet will *not* destroy its magnetic properties, there are subtleties associated with this model as well as with the pole model.

What happens is that you have only scraped off a relatively small number of atoms, whose bound currents do not contribute much to the net magnetic moment.) A right-hand rule due to Ampère tells us how the currents flow, for a given magnetic moment.

Align the thumb of your right hand along the magnetic moment, and with that hand grasp the cylinder. Your fingers will then point along the direction of flow.

As noted above, the magnetic field given by the Amperian approach and the Gilbert approach are identical outside all magnets, and become identical within all magnets after the Gilbert "field" is supplemented. It is usually difficult to find the Amperian currents on the surface of a magnet, whereas it is often easier to find the effective poles for the same magnet.

For one end(pole) of a permanent magnet outside a "soft" magnet, the pole picture of the "soft" magnet has it respond with an image pole of opposite sign to the applied pole; one also can find the Amperian currents on the surface of the "soft" magnet.

POLE NAMING CONVENTIONS

The north pole of the magnet is the pole which, when the magnet is freely suspended, points towards the Earth's magnetic north pole in northern Canada. Since opposite poles(north and south) attract whereas like poles(north and north, or south and south) repel, the Earth's present *geographic north* is thus actually its *magnetic south*. Confounding the situation further, the Earth's magnetic field has reversed itself many times in the distant past.

In order to avoid this confusion, the terms *positive* and *negative* poles are sometimes used instead of *north* and *south*, respectively. As a practical matter, in order to tell which pole of a magnet is north and which is south, it is not necessary to use the earth's magnetic field at all. For example, one calibration method would be to compare it to an electromagnet, whose poles can be identified via the right-hand rule.

Descriptions of Magnetic Behaviours

There are many forms of magnetic behaviour, and all materials exhibit at least one of them. Magnets vary both in the permanency of their magnetization, and in the strength and orientation of the magnetic field they create. This section describes, qualitatively, the primary types of magnetic behaviour that materials can show.

The *physics* underlying each of these behaviours is described in the next section below, and can also be found in more detail in their respective articles.

- Most popularly found in paper clips, paramagnetism is exhibited in substances which do not produce fields by themselves, but which, when exposed to a magnetic field, reinforce that field by becoming magnetized themselves, and thus get attracted to that field. A good example for this behaviour can be found in a bucket of nails - if you pick up a single nail, you can expect that other nails will not follow. However, you can apply an intense magnetic field to the bucket, pick up one nail, and find that many will come with it.
- Unscientifically referred to as 'non-magnetic,' diamagnets actually do exhibit some magnetic behaviour - just to very small magnitudes. In fact, diamagnetic materials, when exposed to a magnetic field, will magnetize(slightly) in the *opposite direction,* getting(slightly) *repelled* from the applied field. Superconductors are strongly diamagnetic.
- Ferromagnetic and ferrimagnetic materials are the 'popular' perception of a magnet. These materials can retain their own magnetization; a common example is a traditional refrigerator magnet.(The difference between ferro- and ferrimagnetic materials is related to their microscopic structure, as explained below.)

PHYSICS OF MAGNETIC BEHAVIOURS

Magnetism, at its root, arises from two sources:

- Electric currents, or more generally moving electric charges, create magnetic fields.
- Many particles have nonzero "intrinsic"(or "spin") magnetic moments.(Just as each particle, by its nature, has a certain mass and charge, each has a certain magnetic moment, possibly zero.)

In magnetic materials, the most important sources of magnetization are, more specifically, the electrons' orbital angular motion around the nucleus, and the electrons' intrinsic magnetic moment. The other potential sources of magnetism are much less important: For example, the nuclear magnetic moments of the nuclei in the material are typically thousands of times smaller than the electrons' magnetic moments, so they are negligible in the context of the magnetization of materials.(Nuclear magnetic moments *are* important in other contexts, particularly in Nuclear Magnetic Resonance(NMR) and Magnetic Resonance Imaging(MRI).)

Ordinarily, the countless electrons in a material are arranged such that their magnetic moments(both orbital and intrinsic) cancel out. This is due, to some extent, to electrons combining into pairs with opposite intrinsic magnetic moments(as a result of the Pauli exclusion principle), or combining into "filled subshells" with zero net orbital motion; in both cases, the electron arrangement is so as to exactly cancel the magnetic moments from each electron.

Moreover, even when the electron configuration *is* such that there are unpaired electrons and/or non-filled subshells, it is often the case that the various electrons in the solid will contribute magnetic moments that point in different, random directions, so that the material will not be magnetic.

However, sometimes(either spontaneously, or due to an applied external magnetic field) each of the electron magnetic moments will be, on average, lined up.

Then the material can produce a net total magnetic field, which can potentially be quite strong. The magnetic behaviour of a material depends on its structure(particularly its electron configuration, for the reasons mentioned above), and also on the temperature(at high temperatures, random thermal motion makes it more difficult for the electrons to maintain alignment).

Physics of Paramagnetism

In a paramagnet there are *unpaired electrons*, i.e. atomic or molecular orbitals with exactly one electron in them. While paired electrons are required by the Pauli exclusion principle to have their intrinsic('spin') magnetic moments pointing in opposite directions(summing to zero), an unpaired electron is free to align its magnetic moment in any direction. When an external magnetic field is applied, these magnetic moments will tend to align themselves in the same direction as the applied field, thus reinforcing it.

Physics of Diamagnetism

In a diamagnet, there are no unpaired electrons, so the intrinsic electron magnetic moments cannot produce any bulk effect. In these cases, the magnetization arises from the electrons' orbital motions, which can be understood classically as follows:

When a material is put in a magnetic field, the electrons circling the nucleus will experience, in addition to their Coulomb attraction to the nucleus, a Lorentz force from the magnetic field. Depending on which direction the electron is orbiting, this force may increase the centripetal force on the electrons, pulling them in towards the nucleus, or it may decrease the force, pulling them away from the nucleus.

This effect systematically increases the orbital magnetic moments that were aligned opposite the field, and decreases the ones aligned parallel to the field(in accordance with Lenz's law). This results in a small bulk magnetic moment, with an opposite direction to the applied field.

Note that this description is meant only as an heuristic; a proper understanding requires a quantum-mechanical description. Note that all materials, including paramagnets, undergo this orbital response. However, in a paramagnet, this response is overwhelmed by the much stronger opposing response described above(i.e., alignment of the electrons' intrinsic magnetic moments).

PHYSICS OF FERROMAGNETISM

A ferromagnet, like a paramagnet, has unpaired electrons. However, in *addition* to the electrons' intrinsic magnetic moments wanting to be parallel to *an applied field*, there is also in these materials a tendency for these magnetic moments to want to be parallel to *each other*.

Thus, even when the applied field is removed, the electrons in the material can keep each other continually pointed in the same direction. Every ferromagnet has its own individual temperature, called the Curie temperature, or Curie point, above which it loses its ferromagnetic properties.

This is because the thermal tendency to disorder overwhelms the energy-lowering due to ferromagnetic order.

MAGNETIC DOMAINS

The magnetic moment of atoms in a ferromagnetic material cause them to behave something like tiny permanent magnets. They stick together and align themselves into small regions of more or less uniform alignment called magnetic domains or Weiss domains.

Magnetic domains can be observed with a magnetic force microscope to reveal magnetic domain boundaries that resemble white lines in the sketch. There are many scientific experiments that can physically show magnetic fields.

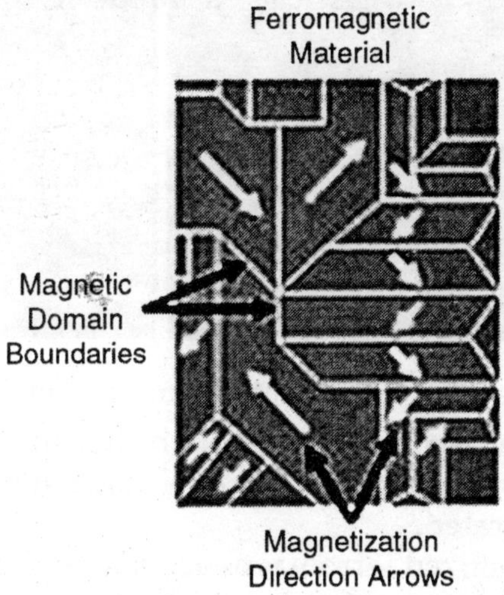

Fig. Magnetic Domains in Ferromagnetic Material

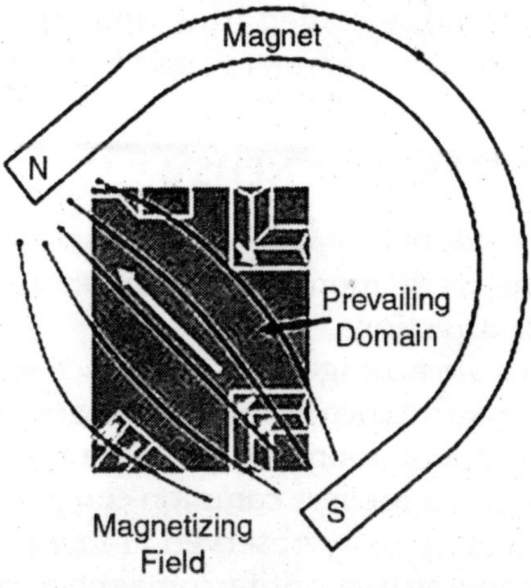

Fig. Effect of a Magnet on the Domains

When a domain contains too many molecules, it becomes unstable and divides into two domains aligned in opposite directions so that they stick together more stably as shown at the right.

When exposed to a magnetic field, the domain boundaries move so that the domains aligned with the magnetic field grow and dominate the structure as shown at the left. When the magnetizing field is removed, the domains may not return to a unmagnetized state. This results in the ferromagnetic material being magnetized, forming a permanent magnet.

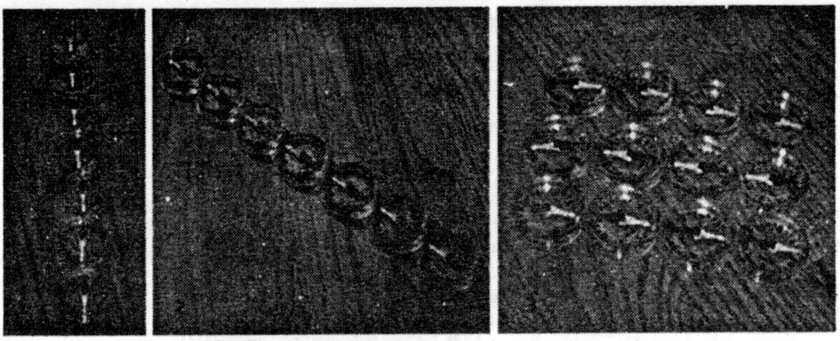

Fig. A Model of Ferromagnetism.

When magnetized strongly enough that the prevailing domain overruns all others to result in only one single domain, the material is magnetically saturated.

When a magnetized ferromagnetic material is heated to the Curie point temperature, the molecules are agitated to the point that the magnetic domains lose the organization and the magnetic properties they cause cease.

When the material is cooled, this domain alignment structure spontaneously returns, in a manner roughly analogous to how a liquid can freeze into a crystalline solid.

PHYSICS OF ANTIFERROMAGNETISM

In an antiferromagnet, unlike a ferromagnet, there is a tendency for the intrinsic magnetic moments of neighboring valence electrons to point in *opposite* directions.

When all atoms are arranged in a substance so that each neighbour is 'anti-aligned', the substance is antiferromagnetic. Antiferromagnets have a zero net magnetic moment, meaning no field is produced by them. Antiferromagnets are less common compared to the other types of behaviours, and are mostly observed at low temperatures.

In varying temperatures, antiferromagnets can be seen to exhibit diamagnetic and ferrimagnetic properties. In some materials, neighboring electrons want to point in opposite directions, but there is no geometrical arrangement in which *each* pair of neighbors is anti-aligned. This is called a spin glass, and is an example of geometrical frustration.

PHYSICS OF FERRIMAGNETISM

Like ferromagnetism, ferrimagnets retain their magnetization in the absence of a field. However, like antiferromagnets, neighboring pairs of electron spins like to point in opposite directions. These two properties are not contradictory, due to the fact that in the optimal geometrical arrangement, there is more magnetic moment from the sublattice of electrons which point in one direction, than from the sublattice which points in the opposite direction.

The first discovered magnetic substance, magnetite, was originally believed to be a ferromagnet; Louis Néel disproved this, however, with the discovery of ferrimagnetism.

OTHER TYPES OF MAGNETISM

There are various other types of magnetism, such as and spin glass, superparamagnetism, superdiamagnetism, and metamagnetism.

COMMON USES OF MAGNETS

Hard disks record data on a thin magnetic coating.

- Magnetic recording media: VHS tapes contain a reel of magnetic tape. The information that makes up the video and sound is encoded on the magnetic coating on the tape. Common audio cassettes also rely on magnetic tape. Similarly, in computers, floppy disks and hard disks record data on a thin magnetic coating.
- Credit, debit, and ATM cards: All of these cards have a magnetic strip on one side. This strip encodes the information to contact an individual's financial institution and connect with their account(s).
- Common televisions and computer monitors: TV and computer screens containing a cathode ray tube employ an electromagnet to guide electrons to the screen. Plasma screens and LCDs use different technologies.
- Speakers and Microphones: Most speakers employ a permanent magnet and a current-carrying coil to convert electric energy(the signal) into mechanical energy(movement which creates the sound). The coil is wrapped around a bobbin attached to the speaker cone, and carries the signal as changing current which

interacts with the field of the permanent magnet. The voice coil feels a magnetic force and in response moves the cone and pressurizes the neighboring air, thus generating sound. Dynamic microphones employ the same concept, but in reverse. A microphone has a diaphragm or membrane attached to a coil of wire. The coil rests inside a specially shaped magnet. When sound vibrates the membrane, the coil is vibrated as well. As the coil moves through the magnetic field, a voltage is induced across the coil. This voltage drives a current in the wire that is characteristic of the original sound.

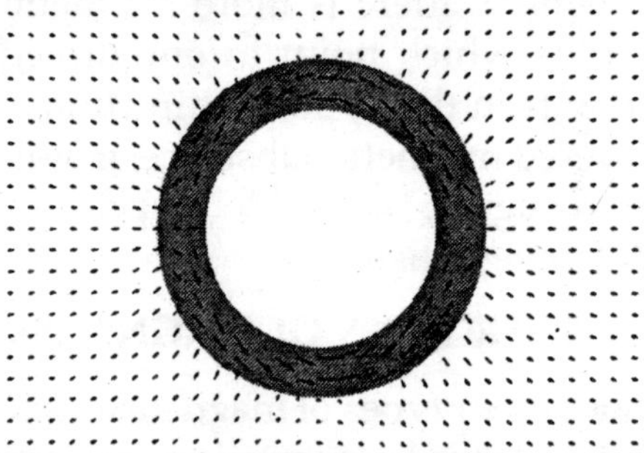

Fig. A Hollow Sphere with μ/μ_0=10, is Immersed in a Uniform, Externally Imposed Magnetic Field. The Interior of the Sphere is Shielded from the Field. The Arrows Map the Magnetic Field B.

Magnetic Hand Separator for Heavy Minerals

- Electric motors and generators: Some electric motors(much like loudspeakers) rely upon a combination of an electromagnet and a permanent magnet, and much like loudspeakers, they convert electric energy into mechanical energy. A generator is the reverse: it converts mechanical energy into electric energy by moving a conductor through a magnetic field.
- Transformers: Transformers are devices that transfer electric energy between two windings of wire that are electrically isolated but are coupled magnetically.
- Chucks: Chucks are used in the metalworking field to hold objects. Magnets are also used in other types of fastening devices, such as the magnetic base, the magnetic clamp and the refrigerator magnet.
- Compasses: A compass(or mariner's compass) is a magnetized pointer free to align itself with a magnetic field, most commonly Earth's magnetic field.
- Art: Vinyl magnet sheets may be attached to paintings,

photographs, and other ornamental articles, allowing them to be attached to refrigerators and other metal surfaces.

- Science Projects: Many topic questions are based on magnets. For example: how is the strength of a magnet affected by glass, plastic, and cardboard?
- Magnets have many uses in toys. M-tic uses magnetic rods connected to metal spheres for construction
- Toys: Due to their ability to counteract the force of gravity at close range, magnets are often employed in children's toys such as the Magnet Space Wheel to amusing effect.
- Magnets can be used to make jewelry. Necklaces and bracelets can have a magnetic clasp, or may be constructed entirely from a linked series of magnets and ferrous beads.
- Magnets can pick up magnetic items(iron nails, staples, tacks, paper clips) that are either too small, too hard to reach, or too thin for fingers to hold. Some screwdrivers are magnetized for this purpose.
- Magnets can be used in scrap and salvage operations to separate magnetic metals(iron, steel, and nickel) from non-magnetic metals(aluminum, non-ferrous alloys, *etc.*). The same idea can be used in the so-called "magnet test", in which an auto body is inspected with a magnet to detect areas repaired using fiberglass or plastic putty.
- Magnetic levitation transport, or maglev, is a form of transportation that suspends, guides and propels vehicles(especially trains) via electromagnetic force. The maximum recorded speed of a maglev train is 581 kilometres per hour(361 mph)
- Magnets may be used to connect some cables to serve as a fail-safe if the cord is pulled.

Magnetization and Demagnetization

Ferromagnetic materials can be magnetized in the following ways:

- Placing the item in an external magnetic field will result in the item retaining some of the magnetism on removal. Vibration has been shown to increase the effect. Ferrous materials aligned with the earth's magnetic field and which are subject to vibration(e.g. frame of a conveyor) have been shown to acquire significant residual magnetism. A magnetic field much stronger than the earth's can be generated inside a solenoid by passing direct current through it.
- Stroking - An existing magnet is moved from one end of the item to the other repeatedly in the same direction.

- Placing a steel bar in a magnetic field, then heating it to a high temperature and then finally hammering it as it cools. This can be done by laying the magnet in a North-South direction in the Earth's magnetic field. In this case, the magnet is not very strong but the effect is permanent.

Permanent magnets can be demagnetized in the following ways:

- Heating a magnet past its Curie point will destroy the long range ordering.
- Contact through stroking one magnet with another in random fashion will demagnetize the magnet being stroked, in some cases; some materials have a very high coercive field and cannot be demagnetized with other permanent magnets.
- Hammering or jarring will destroy the long range ordering within the magnet.
- A magnet being placed in a solenoid which has an alternating current being passed through it will have its long range ordering disrupted, in much the same way that direct current can cause ordering.

In an electromagnet which uses a soft iron core, ceasing the current will eliminate the magnetic field. However, a slight field may remain in the core material as a result of hysteresis.

TYPES OF PERMANENT MAGNETS

MAGNETIC METALLIC ELEMENTS

Many materials have unpaired electron spins, and the majority of these materials are paramagnetic. When the spins interact with each other in such a way that the spins align spontaneously, the materials are called ferromagnetic(what is often loosely termed as "magnetic").

Due to the way their regular crystalline atomic structure causes their spins to interact, some metals are(ferro)magnetic when found in their natural states, as ores. These include iron ore(magnetite or lodestone), cobalt and nickel, as well the rare earth metals gadolinium and dysprosium(when at a very low temperature). Such naturally occurring(ferro)magnets were used in the first experiments with magnetism. Technology has since expanded the availability of magnetic materials to include various manmade products, all based, however, on naturally magnetic elements.

COMPOSITES

Ceramic or Ferrite

Ceramic, or ferrite, magnets are made of a sintered composite of powdered iron oxide and barium/strontium carbonate ceramic. Due to the low cost of the materials and manufacturing methods, inexpensive magnets(or nonmagnetized ferromagnetic cores, for use in electronic component such as radio antennas, for example) of various shapes can be easily mass produced. The resulting magnets are noncorroding, but brittle and must be treated like other ceramics.

Alnico

Alnico magnets are made by casting or sintering a combination of aluminium, nickel and cobalt with iron and small amounts of other elements added to enhance the properties of the magnet. Sintering offers superior mechanical characteristics, whereas casting delivers higher magnetic fields and allows for the design of intricate shapes. Alnico magnets resist corrosion and have physical properties more forgiving than ferrite, but not quite as desirable as a metal.

Ticonal

Ticonal magnets are an alloy of titanium, cobalt, nickel, and aluminum, with iron and small amounts of other elements. It was developed by Philips for loudspeakers.

Injection Molded

Injection molded magnets are a composite of various types of resin and magnetic powders, allowing parts of complex shapes to be manufactured by injection molding. The physical and magnetic properties of the product depend on the raw materials, but are generally lower in magnetic strength and resemble plastics in their physical properties.

Flexible

Flexible magnets are similar to injection molded magnets, using a flexible resin or binder such as vinyl, and produced in flat strips, shapes or sheets. These magnets are lower in magnetic strength but can be very flexible, depending on the binder used. Flexible magnets can be used in industrial printers.

Rare Earth Magnets

'Rare earth'(lanthanoid) elements have a partially occupied *f* electron shell(which can accommodate up to 14 electrons.)

The spin of these electrons can be aligned, resulting in very strong magnetic fields, and therefore these elements are used in compact high-strength magnets where their higher price is not a concern. The most common types of rare earth magnets are samarium-cobalt and neodymium-iron-boron(NIB) magnets.

SINGLE-MOLECULE MAGNETS(SMMS) AND SINGLE-CHAIN MAGNETS(SCMS)

In the 1990s it was discovered that certain molecules containing paramagnetic metal ions are capable of storing a magnetic moment at very low temperatures.

These are very different from conventional magnets that store information at a "domain" level and theoretically could provide a far denser storage medium than conventional magnets. In this direction research on monolayers of SMMs is currently under way. Very briefly, the two main attributes of an SMM are:

- A large ground state spin value(S), which is provided by ferromagnetic or ferrimagnetic coupling between the paramagnetic metal centres.
- A negative value of the anisotropy of the zero field splitting(D)

Most SMM's contain manganese, but can also be found with vanadium, iron, nickel and cobalt clusters. More recently it has been found that some chain systems can also display a magnetization which persists for long times at relatively higher temperatures. These systems have been called single-chain magnets.

Nano-Structured Magnets

Some nano-structured materials exhibit energy waves called magnons that coalesce into a common ground state in the manner of a Bose-Einstein condensate.

Costs

The current cheapest permanent magnets, allowing for field strengths, are flexible and ceramic magnets, but these are also among the weakest types.

Neodymium-iron-boron(NIB) magnets are among the strongest. These cost more per kilogram than most other magnetic materials, but due to their intense field, are smaller and cheaper in many applications.

Temperature

Temperature sensitivity varies, but when a magnet is heated to a temperature known as the Curie point, it loses all of its magnetism, even after cooling below that temperature. The magnets can often be remagnetised however. Additionally some magnets are brittle and can fracture at high temperatures.

ELECTROMAGNETS

An electromagnet in its simplest form, is a wire that has been coiled into one or more loops, known as a solenoid. When electric current flows through the wire, a magnetic field is generated. It is concentrated near(and especially inside) the coil, and its field lines are very similar to those for a magnet. The orientation of this effective magnet is determined via the right hand rule.

The magnetic moment and the magnetic field of the electromagnet are proportional to the number of loops of wire, to the cross-section of each loop, and to the current passing through the wire. If the coil of wire is wrapped around a material with no special magnetic properties(e.g., cardboard), it will tend to generate a very weak field.

However, if it is wrapped around a "soft" ferromagnetic material, such as an iron nail, then the net field produced can result in a several hundred- to thousandfold increase of field strength. Uses for electromagnets include particle accelerators, electric motors, junkyard cranes, and magnetic resonance imaging machines.

Some applications involve configurations more than a simple magnetic dipole, for example quadrupole and sextupole magnets are used to focus particle beams.

UNITS AND CALCULATIONS IN MAGNETISM

How we write the laws of magnetism depends on which set of units we employ. For most engineering applications, MKS or SI(Système International) is common.

Two other sets, Gaussian and CGS-emu, are the same for magnetic properties, and are commonly used in physics. In all units it is convenient to employ two types of magnetic field, B and H, as well as the magnetization *M*, defined as the magnetic moment per unit volume.

- The magnetic induction field B is given in SI units of teslas(T). B is the true magnetic field, whose time-variation produces, by Faraday's Law, circulating electric fields(which the power companies sell). B also produces a deflection force on moving charged particles(as in TV tubes). The tesla is equivalent to the magnetic flux(in webers) per unit area(in meters squared), thus giving B the unit of a flux density. In CGS the unit of B is the gauss(G). One tesla equals 10 G.
- The magnetic field H is given in SI units of ampere-turns per meter(A-turn/m). The "turns" appears because when H is produced by a current-carrying wire, its value is proportional to the number of turns of that wire. In CGS the unit of H is the oersted(Oe). One A-turn/m equals 4ð x 10⁻ Oe.

- The magnetization M is given in SI units of amperes per meter(A/m). In CGS the unit of M is the emu, or electromagnetic unit. One A/m equals 10^{-} emu. A good permanent magnet can have a magnetization as large as a million amperes per meter. Magnetic fields produced by current-carrying wires would require comparably huge currents per unit length, one reason we employ permanent magnets and electromagnets.
- In SI units, the relation $B = \mu_0(H + M)$ holds, where μ_0 is the permeability of space, which equals $4\mu \times 10^{-}$ tesla meters per ampere. In CGS it is written as $B = H + 4\pi M$. [The pole approach gives $\mu_0 H$ in SI units. A $\mu_0 M$ term in SI must then supplement this $\mu_0 H$ to give the correct field within B the magnet. It will agree with the field B calculated using Amperian currents.]

Materials that are not permanent magnets usually satisfy the relation $M = \chi H$ in SI, where χ is the(dimensionless) magnetic susceptibility. Most non-magnetic materials have a relatively small χ(on the order of a millionth), but soft magnets can have ÷ on the order of hundreds or thousands.

For materials satisfying $M = \chi H$, we can also write $B = \mu_0(1 + \chi)H = \mu_0\mu_r H = \mu H$, where $\mu_r = 1 + \chi$ is the(dimensionless) relative permeability and $\mu = \mu_0\mu_r$ is the magnetic permeability. Both hard and soft magnets have a more complex, history-dependent, behaviour described by what are called hysteresis loops, which give either B vs H or M vs H. In CGS $M = \chi H$, but $\chi_{SI} = 4\pi\chi_{CGS}$, and $\mu = \mu_r$.

Caution: In part because there are not enough Roman and Greek symbols, there is no commonly agreed upon symbol for magnetic pole strength and magnetic moment. The symbol m has been used for both pole strength(unit = A·m, where here the upright m is for meter) and for magnetic moment(unit = A·m²).

The symbol μ has been used in some texts for magnetic permeability and in other texts for magnetic moment.

We will use μ for magnetic permeability and m for magnetic moment. For pole strength we will employ q_m. For a bar magnet of cross-section A with uniform magnetization M along its axis, the pole strength is given by $q_m = MA$, so that M can be thought of as a pole strength per unit area.

FIELDS OF A MAGNET

Far away from a magnet, the magnetic field created by that magnet is almost always described(to a good approximation) by a dipole field characterized by its total magnetic moment. This is true regardless of the shape of the magnet, so long as the magnetic moment is nonzero.

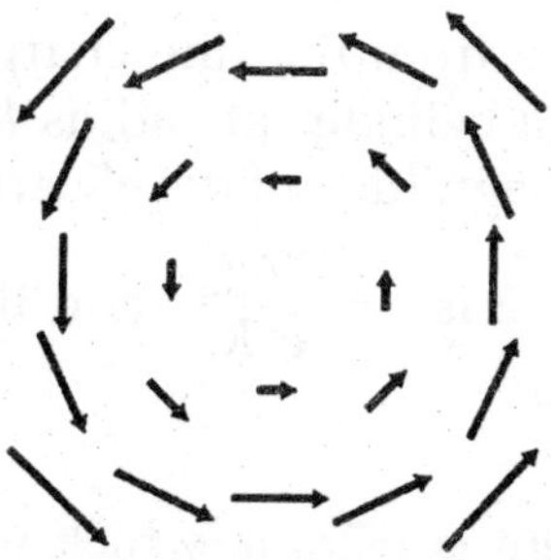

Fig. A Field that Goes in a Circle

One characteristic of a dipole field is that the strength of the field falls off inversely with the cube of the distance from the magnet's centre. Closer to the magnet, the magnetic field becomes more complicated, and more dependent on the detailed shape and magnetization of the magnet. Formally, the field can be expressed as a multipole expansion: A dipole field, plus a quadrupole field, plus an octupole field, etc. At close range, many different fields are possible. For example, for a long, skinny bar magnet with its north pole at one end and south pole at the other, the magnetic field near either end falls off inversely with the square of the distance from that pole.

Calculating the Magnetic Force

Calculating the attractive or repulsive force between two magnets is, in the general case, an extremely complex operation, as it depends on the shape, magnetization, orientation and separation of the magnets.

The pole description is useful to practicing magneticians who design real-world magnets, but real magnets have a pole distribution more complex than a single north and south. Therefore, implementation of the pole idea is not simple. In some cases, one of the more complex formulae given below will be more useful.

Every electron, on account of its spin, is a small magnet. In most materials, the countless electrons have randomly oriented spins, leaving no magnetic effect on average. However, in a bar magnet many of the electron spins are aligned in the same direction, so they act cooperatively, creating a net magnetic field. Let $\pm\gamma$be the charge per unit length of each line charge without relativistic contraction, i.e. in the frame moving with that line charge. Using the approximation $\gamma=(1-v^2/c^2)^{-1/2}\approx 1+v^2/2c^2$ for $v<< c$, the total charge per unit length in frame 2 is

$$\lambda_{total,2} \approx \lambda\left[1+\frac{(u-v)^2}{2c^2}\right]-\lambda\left[1+\frac{(-u-v)}{2c^2}\right]$$

$$=\frac{-2\lambda uv}{c^2}.$$

Let R be the distance from the line charge to the lone charge. Applying Gauss' law to a cylinder of radius R centered on the line charge, we find that the magnitude of the electric field experienced by the lone charge in frame 2 is $E = \frac{4k\lambda uv}{c^2 R}$, and the force acting on the lone charge q is $F = \frac{4k\lambda quv}{c^2 R}$.

In frame 1, the current is $I=2\lambda_1 u$, which we can approximate as $I=2\lambda u$, since the current, unlike $\lambda_{total,\,2}$, doesn't vanish completely without the relativistic effect. The magnetic force on the lone charge q due to the current I is $F = \frac{2kqv}{c^2 R}$.

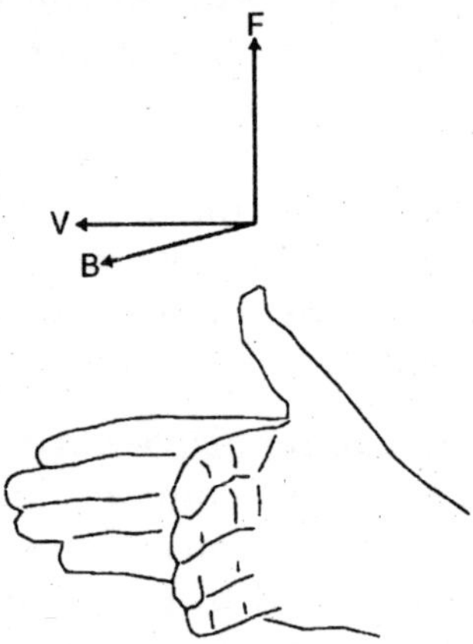

Fig. The Right-Hand Relationship Between the Velocity of a Positively Charged Particle, the Magnetic Field Through Which it is Moving, and the Magnetic Force on it.

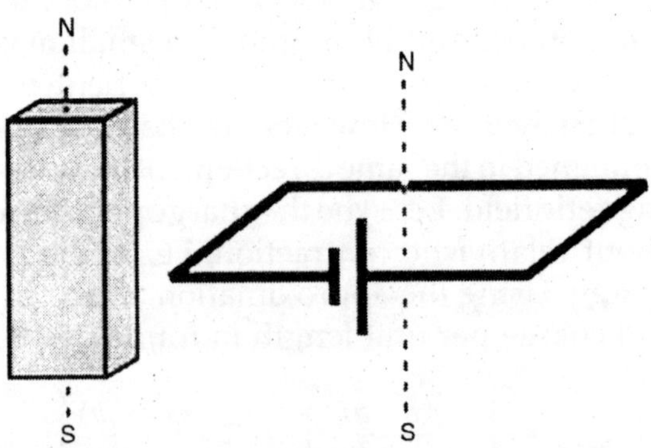

Fig. A Standard Dipole Made from a Square Loop of Wire Shorting Across a Battery. It Acts Very Much Like a bar Magnet, but its Strength is more Easily Quantified.

In addition to the electron's intrinsic magnetic field, there is sometimes an additional magnetic field that results from the electron's orbital motion about the nucleus. This effect is analogous to how a current-carrying loop of wire generates a magnetic field. Again, ordinarily, the motion of the electrons is such that there is no average field from the material, but in certain conditions, the motion can line up so as to produce a measurable total field.

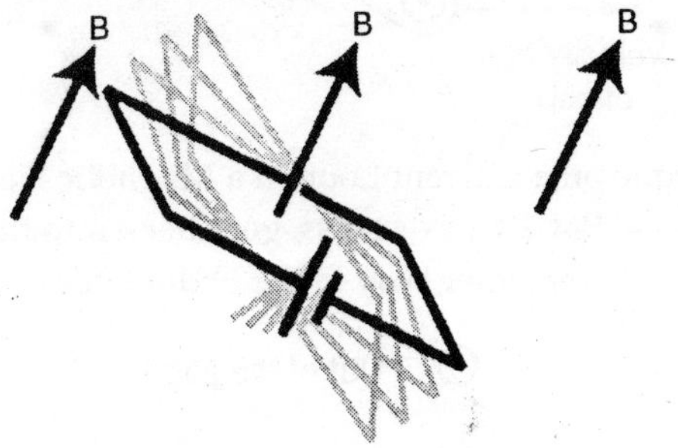

Fig. A Dipole Tends to Align Itself to the Surrounding Magnetic Field.

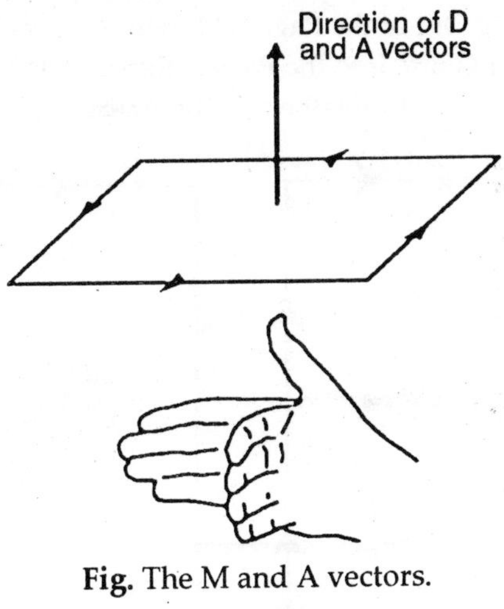

Fig. The M and A vectors.

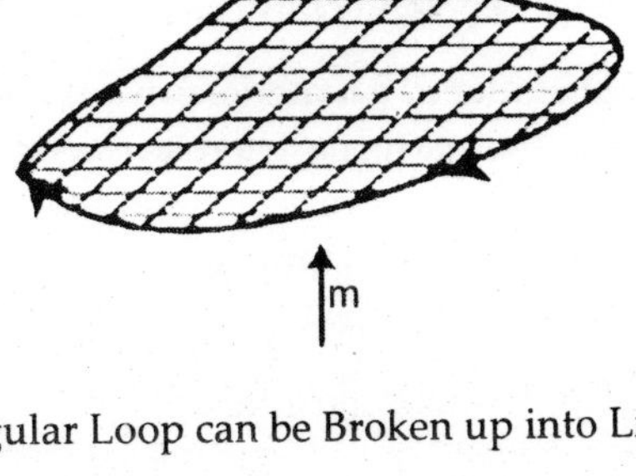

Fig. An Irregular Loop can be Broken up into Little Squares.

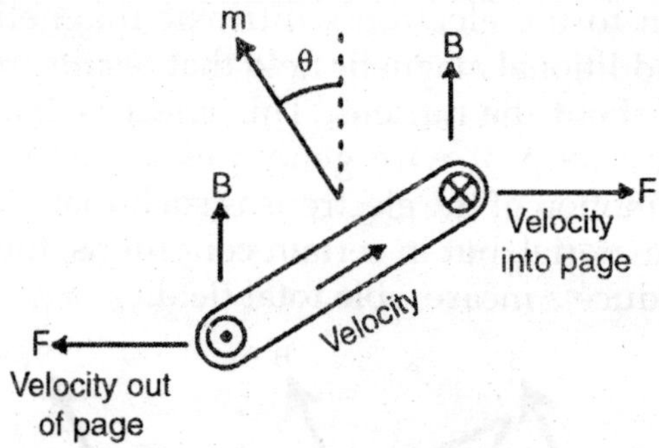

Fig. The Torque on a Current Loop in a Magnetic Field. The Current Comes out of the Page, Goes Across, goes back into the Page, and then back Across the other Way in the Hidden Side of the Loop.

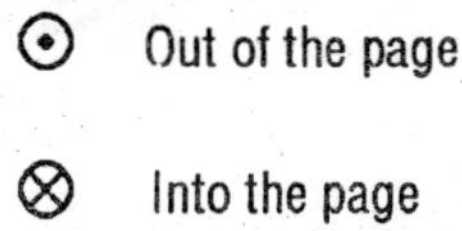

Fig. A Vector Coming Out of the Page is Shown with the Tip of an Arrowhead. A Vector Going into the Page is Represented using the Tailfeathers of the Arrow.

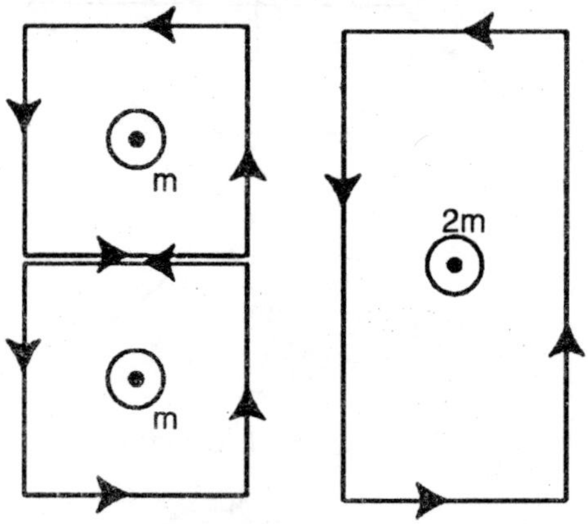

Fig. Dipole Vectors can be Added.

The overall magnetic behaviour of a material can vary widely, depending on the structure of the material, and particularly on its electron configuration. Several forms of magnetic behaviour have been observed in different materials, including:

- Diamagnetism
- Paramagnetism
  - Molecular magnet

- Ferromagnetism
  - Antiferromagnetism
  - Ferrimagnetism
  - Metamagnetism
- Spin glass
- Superparamagnetism

MAGNETISM, ELECTRICITY, AND SPECIAL RELATIVITY

As a consequence of Einstein's theory of special relativity, electricity and magnetism are understood to be fundamentally interlinked. Both magnetism lacking electricity, and electricity without magnetism, are inconsistent with special relativity, due to such effects as length contraction, time dilation, and the fact that the magnetic force is velocity-dependent.

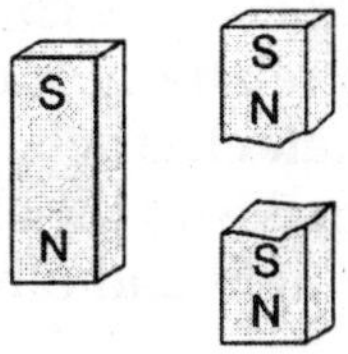

Fig. You Can't Isolate the Poles of a Magnet by Breaking it in Half.

However, when both electricity and magnetism are taken into account, the resulting theory(electromagnetism) is fully consistent with special relativity.

In particular, a phenomenon that appears purely electric to one observer may be purely magnetic to another, or more generally the relative contributions of electricity and magnetism are dependent on the frame of reference. Thus, special relativity "mixes" electricity and magnetism into a single, inseparable phenomenon called electromagnetism(analogously to how special relativity "mixes" space and time into spacetime).

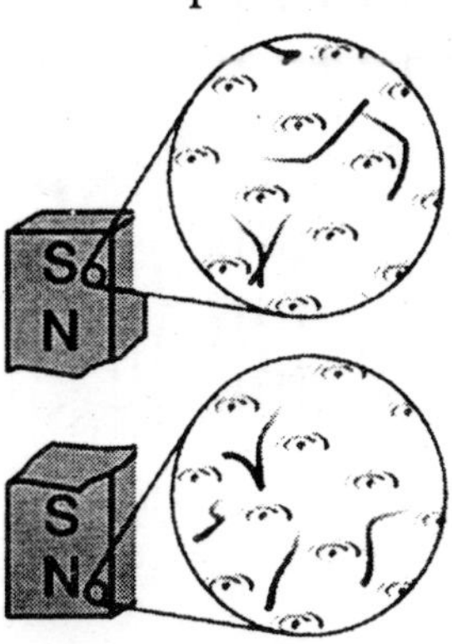

Fig. A Magnetic Dipole is made out of other Dipoles, not out of Monopoles.

Magnetic Fields and Forces

In physics, a *magnetic field* is a vector field that permeates space and which can exert a *magnetic force* on moving electric charges and on magnetic dipoles(such as permanent magnets). When placed in a magnetic field, magnetic dipoles tend to align their axes to be parallel with the magnetic field, as can be seen when iron filings are in the presence of a magnet.

In addition, a changing magnetic field can induce an electric field. Magnetic fields surround and are created by electric currents, magnetic dipoles, and changing electric fields. Magnetic fields also have their own energy, with an energy density proportional to the square of the field intensity. There are some notable specific instances of the magnetic field. First, changes in either of these fields can cause("induce") changes in the other, according to Maxwell's equations. Second, according to Einstein's theory of special relativity, a magnetic force in one inertial frame of reference may be an electric force in another, or vice-versa. Together, these two fields make up the electromagnetic field, which is best known for underlying light and other electromagnetic waves.

B and H

There are two quantities that physicists may refer to as the magnetic field, notated. Although the term "magnetic field" was historically reserved with being termed the "magnetic induction", is now understood to be the more fundamental entity.

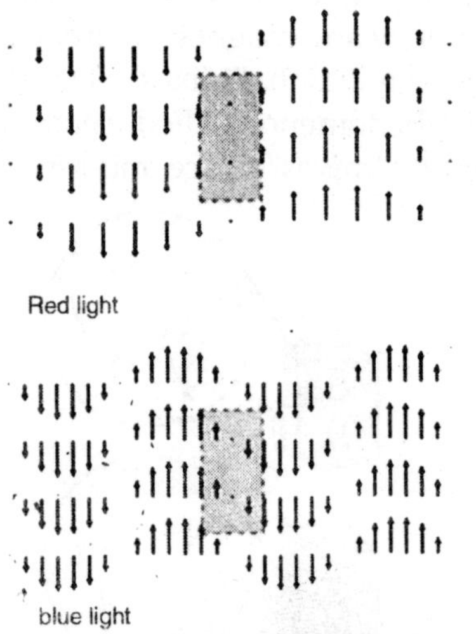

Fig. Red and Blue Light Travel at the Same Speed.

Modern writers vary in their usage of as the magnetic field. A technical paper may fail to make a distinction between the magnetic field and magnetic induction, knowing that the audience may know the difference, but as can be seen in the case of a textbook such as Jackson, the distinction is made precisely.

Alternative Names for B and H

The vector field is known among electrical engineers as the *magnetic field intensity* or *magnetic field strength* and is also known among physicists as *auxiliary magnetic field* or *magnetizing field*. The vector field is known among electrical engineers as *magnetic flux density* or *magnetic induction* or simply *magnetic field*, as used by physicists.

Units

The magnetic field has the SI units of teslas(T), equivalent to webers per square meter(Wb/m^2) or volt seconds per square meter(V s/m^2). In cgs units, has units of gauss(G). The vector field is measured in Amperes/meter(A/m) in SI or oersted(Oe) in cgs units.

Permanent Magnets and Magnetic Poles

The direction of the magnetic field near the poles of a magnet is revealed by placing compasses nearby. As seen here, the magnetic field points towards a magnet's south pole and away from its north pole. Permanent magnets are objects that produce their own persistent magnetic fields.

All permanent magnets have both a north and a south pole.(Magnetic poles always come in north-south pairs.) Like poles repel and opposite poles attract.

The magnetism in a permanent magnet arises from properties of the atoms(in particular the electrons) that compose it. Each atom acts like a little individual magnet. If these magnets line up, they combine to create a macroscopic magnetic effect.

If allowed to twist freely, a magnet will turn to point in the direction of the magnetic field at its location. A compass is a small magnet that uses this effect to point in the direction of the local magnetic field.

By definition, the direction of the magnetic field at a point is the direction that the north pole of a magnet would want to point. If a compass is placed near the north pole of a magnet then it will point away from that pole—like poles repel.

In other words, the magnetic field points away from a magnet near its north pole. The opposite occurs if we place the compass near a magnet's south pole; the magnetic field points towards the magnet near its south pole. Not all magnetic fields are describable in terms of poles, though. A straight current-carrying wire, for instance, produces a magnetic field that points neither towards nor away from the wire, but encircles it instead.

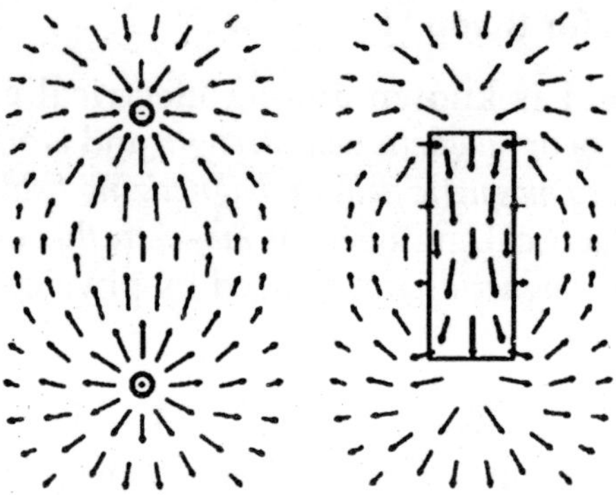

Fig. Magnetic Fields have no Sources or Sinks.

Visualizing the Magnetic Field

The strength and direction of the magnetic field due to an object varies from position to position. Mapping out this magnetic field is simple in principle. First, measure the strength and direction of the magnetic field at a large number of points.

Then mark each location with an arrow(called a vector) pointing in the direction of the magnetic field with a length proportional to the strength of the magnetic field. This is a valid and useful way of marking out and visualizing the magnetic field of an object. It has the unfortunate consequence, though, of cluttering up a graph even when using a small number of points. An alternative method of visualizing the magnetic field is to use "magnetic field lines".

Magnetic B Field Lines

Various physical phenomena have the effect of displaying magnetic field lines. For example, iron filings placed in a magnetic field will line up in such a way as to visually show the orientation of the magnetic field. Another place where magnetic fields are visually displayed is in the polar auroras, in which visible streaks of light line up with the local direction of Earth's magnetic field(due to plasma particle dipole interactions). In these phenomena, lines or curves appear that follow along the direction of the local magnetic field. These field lines provide us with a way to depict or draw the magnetic field(or any other vector field).

Technically, field lines are a set of lines through space whose direction at any point is the direction of the local magnetic field, and whose density is proportional to the magnitude of the local magnetic field. Note that when a magnetic field is depicted with field lines, it is *not* meant to imply that the field is only nonzero along the drawn-in field lines. Rather, the field is typically smooth and continuous everywhere, and can be estimated at *any* point(whether on a field line or not) by looking at the direction and density of the field lines nearby.

The choice of which field lines to draw in such a depiction is arbitrary, apart from the requirement that they be spaced out so that their density approximates the magnitude of the local field. The level of detail at which the magnetic field is depicted can be increased by increasing the number of lines. Field lines are a useful way to represent any vector field and can often be used to reveal sophisticated properties of that field quite simply. One important property of the magnetic field that can be verified with field lines is that it always makes complete loops. Magnetic field lines neither start nor end(although they can extend to or from infinity). To date no exception to this rule has been found.

Even when it appears that a magnetic field has an end(such as when it leaves near a north pole or enters near a south pole of a magnet) in reality it does not. In the case of the permanent magnet the field lines complete the loop inside of the magnet traveling from the south to the north pole.(To see that this must be true imagine placing a compass inside of the magnet. The north pole of the compass will point toward the north pole of the magnet since magnets stacked on each other point in the same direction.) Since magnetic field lines always come in loops, magnetic poles always come in N and S pairs.

If a magnetic field line enters a magnet somewhere it has to leave the magnet somewhere else; it is not allowed to have an end point. For this reason as well, cutting a magnet in half will result in two separate magnets each with both a north and a south pole. Field lines are also a good tool for visualizing magnetic forces. When dealing with magnetic fields in ferromagnetic substances like iron, and in plasmas, the magnetic forces can be understood by imagining that the field lines exert a tension,(like a rubber band) along their length, and a pressure perpendicular to their length on neighboring field lines. The 'unlike' poles of magnets attract because they are linked by many field lines, while 'like' poles repel because the field lines between them don't meet, but run parallel, pushing on each other.

EARTH'S MAGNETIC FIELD

A sketch of Earth's magnetic field representing the source of

Earth's magnetic field as a magnet. The north pole of earth is near the top of the diagram, the south pole near the bottom. Notice that the south pole of that magnet is deep in Earth's interior below Earth's North Magnetic Pole. Earth's magnetic field is produced in the outer liquid part of its core due to a dynamo that produce electrical currents there. Because of Earth's magnetic field, a compass placed anywhere on Earth will turn so that the "north pole" of the magnet inside the compass points roughly north, toward Earth's north magnetic pole in northern Canada.

This is the traditional definition of the "north pole" of a magnet, although other equivalent definitions are also possible. One confusion that arises from this definition is that if Earth itself is considered as a magnet, the *south* pole of that magnet would be the one nearer the north magnetic pole, and vice-versa.(Opposite poles attract and the north pole of the compass magnet is attracted to the north magnetic pole.)

The north magnetic pole is so named not because of the polarity of the field there but because of its geographical location. The figure to the right is a sketch of Earth's magnetic field represented by field lines. The magnetic field at any given point does not point straight toward(or away) from the poles and has a significant up/down component for most locations.(In addition, there is an East/West component as Earth's magnetic poles do not coincide exactly with Earth's geological pole.) The magnetic field is as if there were a magnet deep in Earth's interior. Earth's magnetic field is probably due to a dynamo that produces electric currents in the outer liquid part of its core. Earth's magnetic field is not constant: Its strength and the location of its poles vary. The poles even periodically reverse direction, in a process called geomagnetic reversal.

EFFECTS OF THE MAGNETIC FIELD, B

A magnetic field has many effects on materials and on individual particles. All of these effects can be expressed due to its affects on elementary charges and magnetic dipoles. There are four elementary ways that a magnetic field can affect a charge or a magnetic dipole.

- Sideways force on a moving charge or current
- Torque on a magnetic dipole
- Force on a magnetic dipole due to a non-uniform B
- Force on a charge due to a changing B

Charged particle drifts in a homogenous magnetic field.(A) No disturbing force(B) With an electric field, E(C) With an independent force, F(e.g. gravity)(D) In an inhomogeneous magnetic field, grad H.

FORCE DUE TO A MAGNETIC FIELD ON A MOVING CHARGE

Force on a Charged Particle

A charged particle moving in a magnetic field will feel a *sideways* force that is proportional to the strength of the magnetic field, the component of the velocity that is perpendicular to the magnetic field and the charge of the particle.

This force is known as the Lorentz Force. The force is always perpendicular to both the velocity of the particle and the magnetic field that created it. Neither a stationary particle nor one moving in the direction of the magnetic field lines will experience a force.

For that reason, charged particles move in a circle(or more generally, helix) around magnetic field lines; this is called cyclotron motion. Because the magnetic field is always perpendicular to the motion, the magnetic fields can do no work on a charged particle; a magnetic field alone cannot speed up or slow down a charged particle.

It can and does, however, change the particle's direction, even to the extent that a force applied in one direction can cause the particle to drift in a perpendicular direction.

Force on Current-Carrying Wire

The force on a current carrying wire is similar to that of a moving charge as expected since a charge carrying wire is a collection of moving charges. A current carrying wire will feel a sideways force in the presence of a magnetic field.

The Lorentz force on a macroscopic current is often referred to as the *Laplace force*. The right-hand rule: For a conventional current or moving positive charge in the direction of the thumb of the right hand and the magnetic field along the direction of the fingers(pointing away from palm) the force on the current will be in a direction out of the palm. The direction of the force is reversed for a negative charge.

Direction of Force

The direction of force on a positive charge or a current is determined by the right-hand rule.

Using the right hand and pointing the thumb in the direction of the moving positive charge or positive current and the fingers in the direction of the magnetic field the resulting force on the charge will point outwards from the palm.

The force on a negative charged particle is in the opposite direction. If both the speed and the charge are reversed then the direction of the force remains the same.

For that reason a magnetic field measurement(by itself) cannot distinguish whether there is a positive charge moving to the right or a negative charge moving to the left.(Both of these will produce the same current.) On the other hand, a magnetic field combined with an electric field *can* distinguish between these. An alternative, similar trick to the right hand rule is Fleming's left hand rule.

Torque on a Magnetic Cipole

A magnet placed in a magnetic field will feel a torque that will try to align the magnet with the magnetic field. The torque on a magnet due to an external magnetic field is easy to observe by placing two magnets near each other while allowing one to rotate.

This magnetic torque is the basis for how compasses work. It is used to define the direction of the magnetic field. The magnetic torque also provides the driving torque for simple electric motors.

A magnet(called a rotor) placed on a rotating shaft will feel a strong torque if like poles are placed near its own poles.

If the magnet that caused the rotation—called the stator—is constantly being flipped such that it always has like poles close to the rotor then the rotor will generate a torque that is transferred to the shaft. The polarity of the rotor can easily be flipped if it is an electromagnet by flipping the direction of the current through its coils.

FORCE ON A MAGNETIC DIPOLE DUE TO A NON-UNIFORM B

The most commonly experienced effect of the magnetic field is the force between two magnets: Like poles repel and opposites attract. One can, in fact, express this force in terms of the pole locations and strengths(or more generally, pole distributions) in the two magnets attracting and repulsing each other.

This model is called the "Gilbert model" and produces both the correct force between two magnets, and the correct field outside of the magnets, but the wrong magnetic field *inside* the magnets.(Although the Gilbert model is useful in certain contexts as a mathematical model, the idea of "poles" does not accurately reflect what physically happens inside a magnet.) A more physically accurate picture would be based on the fundamental fact that a magnetic dipole experiences a force, when placed in a *non-uniform* external magnetic field.(In a uniform field, it will experience a torque but no force.)

The south pole of one magnet is attracted to the north pole of another magnet because of the specific way in which each of the microscopic dipoles in either magnet responds to the non-uniform field of the other magnet. The force on a magnetic dipole does not depend directly on the strength or direction of the magnetic field, but only on how these vary with location. A magnet will move to maximize the magnetic field in the direction of its magnetic moment.

Care should be taken to distinguish the magnetic force on a magnetic dipole from the magnetic force on a moving charge. The magnetic force on a charge only occurs when the charge is moving and is in a sideways direction.

It is felt for both uniform and non-uniform magnetic fields. The magnetic force on a dipole, on the other hand, is present only in non-uniform(in space) fields and is in the direction that increases the component of the magnetic field in the direction parallel to the dipole's magnetic moment.

Neither does the force on a magnetic dipole depend on its speed(except at velocities approaching the speed of light).

ELECTRIC FORCE DUE TO A CHANGING B

If the magnetic field in an area is varying with time it generates an electric field that forms closed loops around that area. A conducting wire that forms a closed loop around the area will have an induced voltage generated by this changing magnetic field. This effect is represented mathematically as Faraday's Law and forms the basis of many generators. Care must be taken to understand that the changing magnetic field is a source for an *extended* electric field.

The changing magnetic field does not only create an electric field at that location; rather it generates an electric field that forms closed loops around the location where the magnetic field is changing. Mathematically, Faraday's law is most often represented in terms of the change of magnetic flux with time.

The magnetic flux is the property of a closed loop(say of a coil of wire) and is the product of the area times the magnetic field that is normal to that area.

Engineers and physicists often use magnetic flux as a convenient physical property of a loop(s). They then express the magnetic field as the magnetic flux per unit area. It is for this reason that the field is often referred to as the "magnetic flux density". This approach has the benefit of making certain calculations easier such as in magnetic circuits. It is typically not used outside of electrical circuits, though, because the magnetic field truly is the more 'fundamental' quantity in that it directly connects all of electrodynamics in the simplest manner.

Sources of Magnetic Fields

Magnetic fields can be created in a number of different ways. All of these ways are based on three elementary ways to create a magnetic field.

- Electrical currents(moving charges)
- Magnetic dipoles
- Changing electric field

These sources are thought to affect the virtual particles that compose the field.

Electrical Currents(Moving Charges)

All moving charges produce a magnetic field. The magnetic field of a moving charge is very complicated but is well known. It forms closed loops around a line that is pointing in the direction the charge is moving. The magnetic field of a current on the other hand is much easier to calculate.

Magnetic Field of a Steady Current

Current(*I*) through a wire produces a magnetic field() around the wire. The field is oriented according to the right hand grip rule. The magnetic field generated by a *steady current*(a continual flow of charges, for example through a wire, which is constant in time and in which charge is neither building up nor depleting at any point), is described by the Biot-Savart law. This is a consequence of Ampere's law, one of the four Maxwell's equations that describe electricity and magnetism. The magnetic field lines generated by a current carrying wire form concentric circles around the wire. The direction of the magnetic field of the loops is determined by the right hand grip rule.

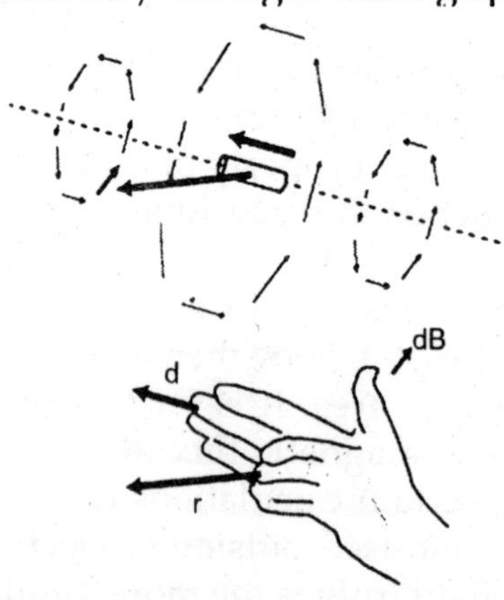

Fig. The Geometry of the Biot-Savart law. The small Arrows Show the Result of the Biot-Savart law at Various Positions Relative to the Current Segment. The Biot-Savart Law Involves a Cross Product, and the right-hand rule for this Cross Product is Demonstrated for one Case.

The strength of the magnetic field decreases with distance from the wire. A current carrying wire can be bent in a loop such that the field is concentrated(and in the same direction) inside of the loop. The field will be weaker outside of the loop. Stacking many such loops to form a solenoid(or long coil) can greatly increase the magnetic field in the centre and decrease the magnetic field outside of the solenoid.

Such devices are called electromagnets and are extremely important in generating strong and well controlled magnetic fields. An infinitely long solenoid will have a uniform magnetic field inside of the loops and no magnetic field outside.

A finite length electromagnet will produce essentially the same magnetic field as a uniform permanent magnet of the same shape and size. An electromagnet has the advantage, though, that you can easily vary the strength(even creating a field in the opposite direction) simply by controlling the input current.

One important use is to continually switch the polarity of a stationary electromagnet to force a rotating permanent magnet to continually rotate using the fact that opposite poles attract and like poles repel. This can be used to create an important type of electrical motor.

Magnetic Dipoles

Magnetic field lines around a "magnetostatic dipole" the magnetic dipole itself is in the centre and is seen from the side. The magnetic field due to a permanent magnet is well known. But, what causes the magnetic field of a permanent magnet? The answer again is that the magnetic field is essentially created due to currents. But this time it is due to the cumulative effect of many small 'currents' of electrons 'orbiting' the nuclei of the magnetic material.

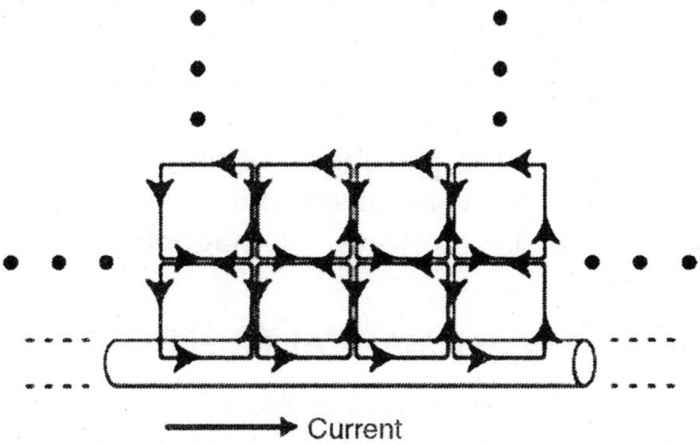

Fig. A Long, Straight Current-Carrying Wire can be Constructed by Filling Half of a Plane with Square Dipoles.

Alternatively it is due to the structure of the electron itself which, in some sense, can be thought of as forming a tiny loop of current.(The true nature of the electron's magnetic field is relativistic in nature, but this model often works.) Both of these tiny loops are modeled in terms of what is called the magnetic dipole. The dipole moment of that dipole can be defined as the current times the area of the loop, then an equation for the magnetic field due to that magnetic dipole can be derived. Magnetic field of a larger magnet can be calculated by adding up the magnetic fields of many magnetic dipoles.

Changing Electric Field

The final known source of magnetic fields is a changing electric field. Just as a changing magnetic field generates an electric field so does a changing electric field generate a magnetic field.(These two effects bootstrap together to form electromagnetic waves, such as light.)

Similar to the way magnetic field lines form close loops around a current a time varying electric field generates a magnetic field that forms closed loops around the region where the electric field is changing. The strength of this magnetic field is proportional to the time rate of the change of the electric field(which is called the displacement current). The fact that a changing electric field creates a magnetic field is known as Maxwell's correction to Ampere's Law.

Magnetic Monopole(Hypothetical)

The magnetic monopole is a hypothetical particle(it may or may not exist). A magnetic monopole would have, as its name suggests, only one pole. In other words, it would possess "magnetic charge" analogous to electric charge.

Positive magnetic charge would correspond to an isolated north pole, and negative magnetic charge would correspond to an isolated south pole. Modern interest in this concept stems from particle theories, notably Grand Unified Theories and superstring theories, that predict either the existence or the possibility of magnetic monopoles. These theories and others have inspired extensive efforts to search for monopoles. Despite these efforts, no magnetic monopole has been observed to date.

DEFINITION AND MATHEMATICAL PROPERTIES OF B

There are several different but physically equivalent ways to define the magnetic field. In principle any of the above effects due to the magnetic field or any of the sources of the magnetic field can be used to define its magnitude and the direction.

Its direction at a given point can be thought of as being the direction that a *hypothetical* freely rotating small test dipole would rotate to point if it *were* placed at that point. Its magnitude is defined(in SI units) in terms of the voltage induced per unit area on a current carrying loop in a uniform magnetic field normal to the loop when the magnetic field is reduced to zero in a unit amount of time.

The SI unit of magnetic field is the Tesla. The magnetic field vector is a pseudovector(also called an axial vector).(This is a technical statement about how the magnetic field behaves when you reflect the world in a mirror.) This fact is apparent from many of the definitions and properties of the field; for example, the magnitude of the field is proportional to the torque on a dipole, and torque is a well-known pseudovector.

Maxwell's Equations

As discussed above, the magnetic field is a vector field.(The magnetic field at each point in space and time is represented by its own vector.) As a vector field, the magnetic field has two important mathematical properties. These properties, along with the corresponding properties of the electric field, make up Maxwell's Equations.

The first is that the magnetic field never starts nor ends at a point. Whatever magnetic field lines enter a region has to eventually leave that region.

This is mathematically equivalent to saying that the divergence of the magnetic is zero.(Such vector fields are called solenoidal vector fields.) The eight original Maxwell's equations can be written in modern vector notation as follows:

(A) The law of total currents

$$J_{tot} = J + \frac{\partial D}{\partial t}$$

(B) The equation of magnetic force

$$\mu H = \Delta \times A$$

(C) Ampère's circuital law

$$\Delta \times H = J_{tot}$$

(D) Electromotive force created by convection, induction, and by static electricity.(This is in effect the Lorentz force)

$$E = \mu v \times H - \frac{\partial A}{\partial t} - \Delta\phi$$

(E) The electric elasticity equation

$$E = \frac{1}{\epsilon} D$$

(F) Ohm's law

$$E = \frac{1}{\sigma} J$$

(G) Gauss' law

$$\Delta \cdot D = \rho$$

(H) Equation of continuity

$$\Delta \cdot J = -\frac{\partial \rho}{\partial t}$$

Notation

H is the magnetizing field, which Maxwell called the "magnetic intensity".

J is the electric current density(with J_{tot} being the total current including displacement current).

D is the displacement field(called the "electric displacement" by Maxwell).

ρ is the free charge density(called the "quantity of free electricity" by Maxwell).

A is the magnetic vector potential(called the "angular impulse" by Maxwell).

E is called the "electromotive force" by Maxwell. The term electromotive force is nowadays used for voltage, but it is clear from the context that Maxwell's meaning corresponded more to the modern term electric field.

Φ is the electric potential(which Maxwell also called "electric potential").

σ is the electrical conductivity(Maxwell called the inverse of conductivity the "specific resistance", what is now called the resistivity).

This property is called Gauss' law for magnetism and is one of Maxwell's Equations.

It is also equivalent to the statement that there are no magnetic monopoles. The second mathematical property of the magnetic field is that it always loops around the source that creates it.

This source could be a current, a magnet, or a changing electric field, but it is always within the loops of magnetic field they create.

Mathematically, this fact is described by the Ampère-Maxwell equation.

MEASURING THE MAGNETIC B FIELD

There are many ways of measuring the magnetic field, many of which use the effects described above. Devices used to measure the local magnetic field are called magnetometers. Important magnetometers include using a rotating coil, Hall effect magnetometers, NMR magnetometer, SQUID magnetometer, and a fluxgate magnetometer. The magnetic fields of distant astronomical objects can be determined by noting their effects on local charged particles.

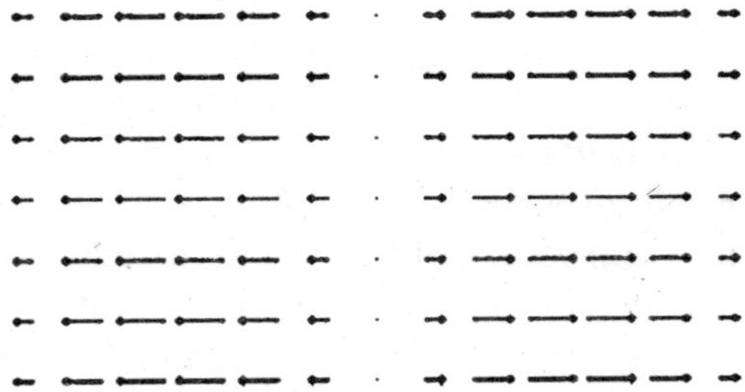

Fig. An Impossible Wave Pattern.

For instance, electrons spiraling around a field line will produce synchotron radiation which is detectable in radio waves.

Hall Effect

Because the Lorentz force is charge-sign-dependent, it results in charge separation when a conductor with current is placed in a transverse magnetic field, with a buildup of opposite charges on two opposite sides of conductor in the direction normal to the magnetic field, and the potential difference between these sides can be measured. The Hall effect is often used to measure the magnitude of a magnetic field as well as to find the sign of the dominant charge carriers in semiconductors(negative electrons or positive holes).

SQUID Magnetometer

Superconductors are materials with both distinctive electric properties(perfect conductivity) and magnetic properties(such as the Meissner effect, in which many superconductors can perfectly expel magnetic fields). Due to these properties, it turns out that loops that incorporate superconducting material and their Josephson junctions can function as very sensitive magnetometers, called SQUIDs.

The H Field

The term 'magnetic field' can also be used to describe the magnetic field. The magnetic field is similar to in that it is a vector field, but its units are often different. In SI units, and are measured in teslas(T) and amperes per meter(A/m), respectively; or, in cgs units, in gauss(G) and oersteds(Oe), respectively. Outside of magnetizable materials, the two fields are identical(apart from possibly a constant conversion factor), but inside a magnetic material they can differ substantially.

PHYSICAL INTERPRETATION OF THE H FIELD

When magnetic materials are present, the total magnetic field is caused by two different types of currents which need to be distinguished: free current and bound current.

Free currents are the ordinary currents in wires and other conductors, that can be controlled and measured. Bound currents are the tiny circular currents inside atoms that are responsible for the magnetization of magnetic materials.

Although the actual source of the magnetic field in electron orbitals of atoms is complex, the magnetic properties of a material can be accounted for by assuming it is divided into tiny blocks, each of which has a current flowing around its outside surface, perpendicular to the magnetic field axis. As an example of bound current consider a uniform permanent bar magnet.

A piece of iron is formed of many tiny regions called magnetic domains, each of which is a magnetic dipole, essentially a tiny loop of current. In a bar magnet, most of these dipoles have their poles lined up, creating a large magnetic field.

If we add up the currents of all these tiny loops we will find that the currents cancel in the interior of the material, but add up along the sides of the bar.(This current loops around the sides and not at the poles.) No one charge makes the complete trip around the magnet(each charge is bound to its tiny loop) but the net effect is exactly equivalent to a real current that flows around the outside surface of the magnet, perpendicular to the magnetic field axis.(If the magnetization is not uniform then a bound current will flow through the bulk of the magnetic material as well.)

The magnetic is useful because it treats these two types of currents differently. The free currents it treats in the normal fashion and therefore has the same form as the magnetic field it would generate.

The magnetic fields treats the field inside of a magnetic material(due to that magnetic material) in a manner similar to the Gilbert model.(By subtracting the magnetization from the B field we are essentially converting the bound current sources to Gilbert-like magnetic charges at the poles.)

Unlike the magnetic, which always forms closed loops, the field due to the magnetic charges flow outward(or inward depending on the sign of the magnetic charge) in both directions from the poles. And while the magnetic field is exactly the same on the outside of the magnetic material for both models the magnetic fields inside are quite different. Putting both sources together we see that the magnetic field is the same as the magnetic field to a multiplicative constant outside of magnetic materials, but is completely different from the magnetic field inside a magnetic material. The advantage of this hybrid field is that these sources are treated so differently that we can often pick out one source from the other.

For example a line integral of the magnetic field in a closed loop will yield the total free current in the loop(and not the bound current). This is unlike the magnetic field where a similar integral will yield the sum of both the free and the bound current.

If one wants to isolate the contribution due to the bound currents then a surface integral of over any closed surface will pick out the 'magnetic charges' at the poles.

Sources of the H Field

Unlike the magnetic field that only has a current source such that the magnetic field loops around currents, the magnetic field has two types of sources.

The first source of magnetic field are the *free* currents for which loop around similar to the way field loops around the total current. The second source of the magnetic field is 'magnetic charges' near the poles of the magnetic material.

USES OF THE H FIELD

Energy Stored in Magnetic Fields

In order to create a magnetic field we need to do work to establish a free current. If we are to ask how much energy does it take to create a specific magnetic field using a particular free current then we need to distinguish between the free and the bound currents.

It is the free current that we are 'pushing' on. The bound currents are freeloaders. They create a magnetic field that the free current has to work against without doing any of the work. If we are to calculate the energy of creating a magnetic field we need to have a way of separating out the free current.

The magnetic cannot be used to determine this free current since does not distinguish between bound and free current. The magnetic field does treat the two sources differently. Therefore it is useful in calculating the energy needed to create a magnetic field with a free current in the presence of magnetic materials.

MAGNETIC CIRCUITS

A second use for is in magnetic circuits where inside a linear material. Here, ì is the permeability of the material. This is similar in form to Ohm's Law, where the current density, ó is is the conductance and is the Electric field. Using this analogy it is straight-forward to calculate the magnetic flux of complicated magnetic field geometries, by using all the available techniques of circuit theory.

History of B and H

The difference between the and the vectors can be traced back to Maxwell's 1855 paper entitled *On Faraday's Lines of Force.* It is later clarified in his concept of a sea of molecular vortices that appears in his 1861 paper On Physical Lines of Force - 1861. Within that context, represented pure vorticity(spin), whereas was a weighted vorticity that was weighted for the density of the vortex sea. Maxwell considered magnetic permeability μ to be a measure of the density of the vortex sea.

The electric current equation can be viewed as a convective current of electric charge that involves linear motion. By analogy, the magnetic equation is an inductive current involving spin. There is no linear motion in the inductive current along the direction of the vector. The magnetic inductive current represents lines of force. In particular, it represents lines of inverse square law force.

ROTATING MAGNETIC FIELDS

The rotating magnetic field is a key principle in the operation of alternating-current motors. A permanent magnet in such a field will rotate so as to maintain its alignment with the external field.

This effect was conceptualized by Nikola Tesla, and later utilised in his, and others', early AC(alternating-current) electric motors. A rotating magnetic field can be constructed using two orthogonal coils with 90 degrees phase difference in their AC currents. However, in practice such a system would be supplied through a three-wire arrangement with unequal currents.

This inequality would cause serious problems in standardization of the conductor size and so, in order to overcome it, three-phase systems are used where the three currents are equal in magnitude and have 120 degrees phase difference.

Three similar coils having mutual geometrical angles of 120 degrees will create the rotating magnetic field in this case. The ability of the three-phase system to create a rotating field, utilized in electric motors, is one of the main reasons why three-phase systems dominate the world's electrical power supply systems.

Because magnets degrade with time, synchronous motors and induction motors use short-circuited rotors(instead of a magnet) following the rotating magnetic field of a multicoiled stator. The short-circuited turns of the rotor develop eddy currents in the rotating field of the stator, and these currents in turn move the rotor by the Lorentz force. In 1882, Nikola Tesla identified the concept of the rotating magnetic field. In 1885, Galileo Ferraris independently researched the concept. In 1888, Tesla gained U.S. Patent 381,968 for his work. Also in 1888, Ferraris published his research in a paper to the *Royal Academy of Sciences* in Turin.

SPECIAL RELATIVITY AND ELECTROMAGNETISM

Magnetic fields played an important role in helping to develop the theory of special relativity.

Moving Magnet and Conductor Problem

Imagine a moving conducting loop that is passing by a stationary magnet. Such a conducting loop will have a current generated in it as it passes through the magnetic field. But why? It is answering this seemingly innocent question that led Albert Einstein to develop his theory of special relativity.

A stationary observer would see an unchanging magnetic field and a moving conducting loop. Since the loop is moving all of the charges that make up the loop are also moving. Each of these charges will have a sideways, Lorentz force, acting on it which generates the current.

Meanwhile, an observer on the moving reference frame would see a *changing* magnetic field and *stationary* charges.(The loop is not moving in this observers reference frame. The magnet is.) This changing magnetic field generates an *electric* field.

The stationary observer claims there is *only* a magnetic field that creates a *magnetic force* on a moving charge. The moving observer claims that there is both a magnetic and an electric field but all of the force is due to the *electric* field. Which is true? Does the electric field exist or not? The answer, according to special relativity, is that both observers are right from their reference frame. A pure magnetic field in one reference can be a mixture of magnetic and electric field in another reference frame.

Electric and Magnetic Fields Different Aspects of the Same Phenomenon

According to special relativity, electric and magnetic forces are part of a single physical phenomenon, electromagnetism; an electric force perceived by one observer will be perceived by another observer in a different frame of reference as a mixture of electric and magnetic forces. A magnetic force can be considered as simply the relativistic part of an electric force when the latter is seen by a moving observer.

More specifically, rather than treating the electric and magnetic fields as separate fields, special relativity shows that they naturally mix together into a rank-2 tensor, called the electromagnetic tensor. This is analogous to the way that special relativity "mixes" space and time into spacetime, and mass, momentum and energy into four-momentum.

MAGNETIC FIELD SHAPE DESCRIPTIONS

Schematic quadrupole magnet(*"four-pole"*) magnetic field. There are four steel pole tips, two opposing magnetic north poles and two opposing magnetic south poles.

- An azimuthal magnetic field is one that runs east-west.
- A meridional magnetic field is one that runs north-south. In the solar dynamo model of the Sun, differential rotation of the solar plasma causes the meridional magnetic field to stretch into an azimuthal magnetic field, a process called the *omega-effect*. The reverse process is called the *alpha-effect*.
- A dipole magnetic field is one seen around a bar magnet or around a charged elementary particle with nonzero spin.
- A quadrupole magnetic field is one seen, for example, between the poles of four bar magnets. The field strength grows linearly with the radial distance from its longitudinal axis.
- A solenoidal magnetic field is similar to a dipole magnetic field, except that a solid bar magnet is replaced by a hollow electromagnetic coil magnet.
- A toroidal magnetic field occurs in a doughnut-shaped coil, the electric current spiraling around the tube-like surface, and is found, for example, in a tokamak.
- A poloidal magnetic field is generated by a current flowing in a ring, and is found, for example, in a tokamak.
- A radial magnetic field is one in which the field lines are directed

from the centre outwards, similar to the spokes in a bicycle wheel. An example can be found in a loudspeaker transducers(driver).
- A helical magnetic field is corkscrew-shaped, and sometimes seen in space plasmas such as the Orion Molecular Cloud.

The phenomenon of magnetism is "mediated" by the magnetic field — i.e., an electric current or magnetic dipole creates a magnetic field, and that field, in turn, imparts magnetic forces on other particles that are in the fields.

To an excellent approximation(but ignoring some quantum effects), Maxwell's equations(which simplify to the Biot-Savart law in the case of steady currents) describe the origin and behaviour of the fields that govern these forces.

Therefore magnetism is seen whenever electrically charged particles are in motion—for example, from movement of electrons in an electric current, or in certain cases from the orbital motion of electrons around an atom's nucleus. They also arise from "intrinsic" magnetic dipoles arising from quantum effects, i.e. from quantum-mechanical spin. The same situations which create magnetic fields(charge moving in a current or in an atom, and intrinsic magnetic dipoles) are also the situations in which a magnetic field has an effect, creating a force. Because this is a cross product, the force is perpendicular to both the motion of the particle and the magnetic field.

It follows that the magnetic force does no work on the particle; it may change the direction of the particle's movement, but it cannot cause it to speed up or slow down.

One tool for determining the direction of the velocity vector of a moving charge, the magnetic field, and the force exerted is labeling the index finger "V", the middle finger "B", and the thumb "F" with your right hand. When making a gun-like configuration(with the middle finger crossing under the index finger), the fingers represent the velocity vector, magnetic field vector, and force vector, respectively.

Lenz's law gives the direction of the induced electromotive force(emf) and current resulting from electromagnetic induction. German physicist Heinrich Lenz formulated it in 1834.

MAGNETIC DIPOLES

In physics, there are two kinds of dipoles(Hellènic: *di(s)-* = two- and *pòla* = pivot, hinge):

- An electric dipole is a separation of positive and negative charge. The simplest example of this is a pair of electric charges of equal magnitude but opposite sign, separated by some, usually small, distance. A permanent electric dipole is called an electret.

- A magnetic dipole is a closed circulation of electric current. A simple example of this is a single loop of wire with some constant current flowing through it.

Dipoles can be characterized by their dipole moment, a vector quantity. For the simple electric dipole given above, the electric dipole moment would point from the negative charge towards the positive charge, and have a magnitude equal to the strength of each charge times the separation between the charges. For the current loop, the magnetic dipole moment would point through the loop(according to the right hand grip rule), with a magnitude equal to the current in the loop times the area of the loop.

In addition to current loops, the electron, among other fundamental particles, is said to have a magnetic dipole moment. This is because it generates a magnetic field which is identical to that generated by a very small current loop. However, to the best of our knowledge, the electron's magnetic moment is not due to a current loop, but is instead an intrinsic property of the electron. It is also possible that the electron has an *electric* dipole moment, although this has not yet been observed.

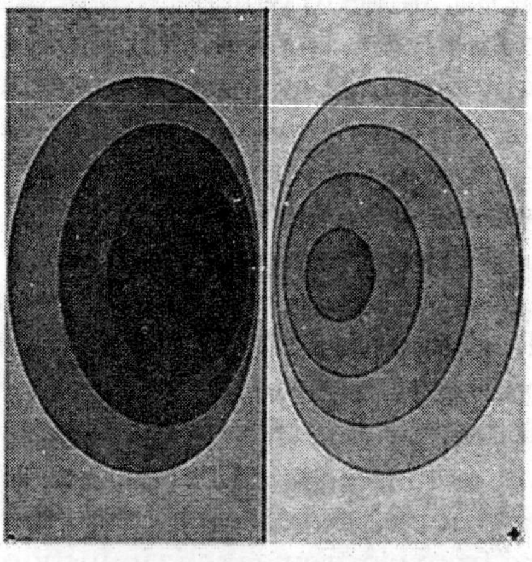

Fig. Contour Plot of an Electrical Dipole, with Equipotential Surfaces Indicated

A permanent magnet, such as a bar magnet, owes its magnetism to the intrinsic magnetic dipole moment of the electron. The two ends of a bar magnet are referred to as poles(not to be confused with monopoles), and are labeled "north" and "south."

The dipole moment of the bar magnet points from its magnetic south to its magnetic north pole—confusingly, the "north" and "south" convention for magnetic dipoles is the opposite of that used to describe the Earth's geographic and magnetic poles, so that the Earth's geomagnetic north pole is the *south* pole of its dipole moment.

This should not be difficult to remember; it simply means that the north pole of a bar magnet is the one which points north if used as a compass. The only known mechanisms for the creation of magnetic dipoles are by current loops or quantum-mechanical spin since the existence of magnetic monopoles has never been experimentally demonstrated.

PHYSICAL DIPOLES, POINT DIPOLES, AND APPROXIMATE DIPOLES

A *physical dipole* consists of two equal and opposite point charges: literally, two poles. Its field at large distances(i.e., distances large in comparison to the separation of the poles) depends almost entirely on the dipole moment as defined above.

A *point(electric) dipole* is the limit obtained by letting the separation tend to 0 while keeping the dipole moment fixed. The field of a point dipole has a particularly simple form, and the order-1 term in the multipole expansion is precisely the point dipole field. Although there are no known magnetic monopoles in nature, there are magnetic dipoles in the form of the quantum-mechanical spin associated with particles such as electrons(although the accurate description of such effects falls outside of classical electromagnetism).

A theoretical magnetic *point dipole* has a magnetic field of the exact same form as the electric field of an electric point dipole. A very small current-carrying loop is approximately a magnetic point dipole; the magnetic dipole moment of such a loop is the product of the current flowing in the loop and the(vector) area of the loop.

Any configuration of charges or currents has a 'dipole moment', which describes the dipole whose field is the best approximation, at large distances, to that of the given configuration.

This is simply one term in the multipole expansion; when the charge("monopole moment") is 0 — as it *always* is for the magnetic case, since there are no magnetic monopoles — the dipole term is the dominant one at large distances: its field falls off in proportion to $1/r$, as compared to $1/r$ for the next(quadrupole) term and higher powers of $1/r$ for higher terms, or $1/r$ for the monopole term.

MOLECULAR DIPOLES

Many molecules have such dipole moments due to non-uniform distributions of positive and negative charges on the various atoms.

For example: A molecule with a permanent dipole moment is called a polar molecule.

A molecule is polarized when it carries an induced dipole. The physical chemist Peter J. W. Debye was the first scientist to study molecular dipoles extensively, and dipole moments are consequently measured in units named *debye* in his honour.

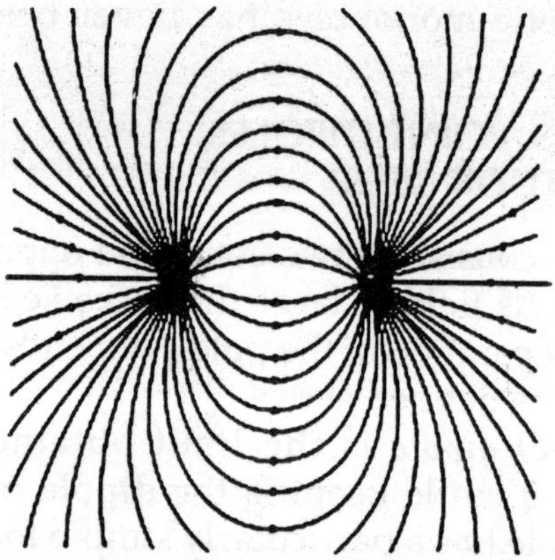

Fig. Electric Dipole Field Lines(positive) H-Cl(negative)

With respect to molecules there are three types of dipoles:

- Permanent dipoles: These occur when two atoms in a molecule have substantially different electronegativity—one atom attracts electrons more than another becoming more negative, while the other atom becomes more positive.
- Instantaneous dipoles: These occur due to chance when electrons happen to be more concentrated in one place than another in a molecule, creating a temporary dipole.
- Induced dipoles These occur when one molecule with a permanent dipole repels another molecule's electrons, "inducing" a dipole moment in that molecule.

The definition of an induced dipole given in the previous sentence is too restrictive and misleading. An induced dipole of *any* polarizable charge distribution ρ(remember that a molecule has a charge distribution) is caused by an electric field external to ρ.

This field may, for instance, originate from an ion or polar molecule in the vicinity of ñ or may be macroscopic(e.g., a molecule between the plates of a charged capacitor). The size of the induced dipole is equal to the product of the strength of the external field and the dipole polarizability of ρ.

Typical gas phase values of some chemical compounds in debye units:

- Carbon dioxide: 0

- Carbon monoxide: 0.112
- Ozone: 0.53
- Phosgene: 1.17
- Water vapour: 1.85
- Hydrogen cyanide: 2.98
- Cyanamide: 4.27
- Potassium bromide: 10.41

These values can be obtained from measurement of the dielectric constant. When the symmetry of a molecule cancels out a net dipole moment, the value is set at 0. The highest dipole moments are in the range of 10 to 11. From the dipole moment information can be deduced about the molecular geometry of the molecule. For example the data illustrate that carbon dioxide is a linear molecule but ozone is not.

QUANTUM MECHANICAL DIPOLE OPERATOR

Consider a collection of *N* particles with charges q_i and position vectors r_i. For instance, this collection may be a molecule consisting of electrons, all with charge *-e*, and nuclei with charge eZ_i, where Z_i is the atomic number of the ith nucleus. The physical quantity(observable) dipole has the quantum mechanical operator:

$$p = \sum_{i=1}^{N} q_i r_i .$$

ATOMIC DIPOLES

A non-degenerate(S-state) atom can have only a zero permanent dipole. This fact follows quantum mechanically from the inversion symmetry of atoms. All 3 components of the dipole operator are antisymmetric under inversion with respect to the nucleus,

$$\Im p \Im^{-1} = -p,$$

Where p is the dipole operator and $\Im$ is the inversion operator. The permanent dipole moment of an atom in a non-degenerate state is given as the expectation(average) value of the dipole operator,

$$\langle p \rangle = \langle S|p|S \rangle,$$

where is $|S\rangle$ an S-state, non-degenerate, wavefunction, which is symmetric or antisymmetric under inversion: $\Im|S\rangle = \pm|S\rangle$. Since the product of the wavefunction(in the ket) and its complex conjugate(in the bra) is always symmetric under inversion and its inverse,

$$\langle p \rangle = \langle \Im^{-1}S|p|\Im^{-1}S \rangle = \langle S|\Im p \Im^{-1}|S \rangle = -\langle p \rangle$$

it follows that the expectation value changes sign under inversion. We used here the fact that $\Im$, being a symmetry operator, is unitary:$\Im^{-1}=\Im^*$and by definition the Hermitian adjoint $\Im^*$ may be moved from bra to ket and then becomes $\Im^{**}=\Im$. Since the only quantity that is equal to minus itself is the zero, the expectation value vanishes, $\langle p\rangle = 0$.

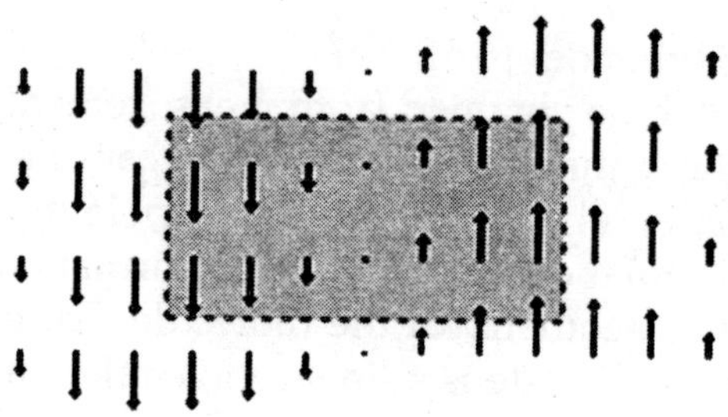

Fig. The Wave Pattern is Curly. For Example, the Circulation Around this Rectangle is Nonzero and Counterclockwise.

In the case of open-shell atoms with degenerate energy levels, one could define a dipole moment by the aid of the first-order Stark effect. This only gives a non-vanishing dipole(by definition proportional to a non-vanishing first-order Stark shift) if some of the wavefunctions belonging to the degenerate energies have opposite parity; i.e., have different behaviour under inversion.

This is a rare occurrence, but happens for the excited H-atom, where 2s and 2p states are "accidentally" and have opposite parity(2s is even and 2p is odd).

FIELD FROM A MAGNETIC DIPOLE

Magnitude

The far-field strength, B, of a dipole magnetic field is given by

$$B(m,r,\lambda) = \frac{\mu_0 m}{4\pi r^3}\sqrt{1+3\sin^2\lambda}$$

where

B is the strength of the field, measured in teslas;

r is the distance from the centre, measured in metres;

λ is the magnetic latitude(90°-θ) where θ = magnetic colatitude, measured in radians or degrees from the dipole axis(Magnetic colatitude is 0 along the dipole's axis and 90° in the plane perpendicular to its axis.); m is the dipole moment(VADM=virtual axial dipole moment), measured in ampere square-metres(A•m), which equals joules per tesla; μ_0 is the permeability of free space, measured in henrys per metre.

Conversion to cylindrical coordinates is achieved using

$r = z + \rho$

and

$$\lambda\text{-arcsin}\left(\frac{z}{\sqrt{z^2+\rho^2}}\right)$$

where ρ is the perpendicular distance from the z-axis. Then,

$$B(\rho, z) = \frac{\mu_0 m}{4\pi\left(z^2+\rho^2\right)^{3/2}}\sqrt{1+\frac{3z^2}{z^2+\rho^2}}$$

Vector form

The field itself is a vector quantity:

$$\mathrm{B}(\mathrm{m,r}) = \frac{\mu_0}{4\pi r^3}(3(\mathrm{m}\cdot\hat{\mathrm{r}})\hat{\mathrm{r}} - \mathrm{m}) + \frac{2\mu_0}{3} m\delta(\mathrm{r})$$

where

B is the field;

r is the vector from the position of the dipole to the position where the field is being measured;

r is the absolute value of r: the distance from the dipole;

r = r/r is the unit vector parallel to r;

m is the(vector) dipole moment;

μ_0 is the permeability of free space;

δ is the three-dimensional delta function.(δ^3(r) = 0 except at r =(0,0,0), so this term is ignored in multipole expansion.)

This is *exactly* the field of a point dipole, *exactly* the dipole term in the multipole expansion of an arbitrary field, and *approximately* the field of any dipole-like configuration at large distances.

Magnetic Vector Potential

The vector potential A of a magnetic dipole is

$$\mathrm{A}(\mathrm{r}) = \frac{\mu_0}{4\pi r^2}(\mathrm{m}\times\hat{\mathrm{r}})$$

with the same definitions as above.

Field from an Electric Dipole

The electrostatic potential at position due to an electric dipole at the origin is given by:

$$\Phi(\mathrm{r}) = \frac{1}{4\pi\epsilon_0}\frac{\mathrm{p}\cdot\hat{\mathrm{r}}}{\mathrm{r}^2}$$

where

$\hat{r}$ is a unit vector in the direction of r;

p is the(vector) dipole moment;

ε_0 is the permittivity of free space.

This term appears as the second term in the multipole expansion of an arbitrary electrostatic potential Φ(r).

If the source of Φ(r) is a dipole, as it is assumed here, this term is the only non-vanishing term in the multipole expansion of Φ(r). The electric field from a dipole can be found from the gradient of this potential:

$$E = -\Delta\Phi = \frac{1}{4\pi\epsilon_0}\left(\frac{3(p\cdot\hat{r})\hat{r}-p}{r^3}\right) - \frac{1}{3\epsilon_0} p\delta^3(r)$$

where E is the electric field and δ is the 3-dimensional delta function.(δ^3 (r) = 0 except at r = (0, 0, 0), so this term is ignored in multipole expansion.)

Notice that this is formally identical to the magnetic field of a point magnetic dipole; only a few names have changed.

TORQUE ON A DIPOLE

Since the direction of an electric field is defined as the direction of the force on a positive charge, electric field lines point away from a positive charge and toward a negative charge. When placed in an electric or magnetic field, equal but opposite forces arise on each side of the dipole creating a torque τ:

$$\tau = p \times E$$

for an electric dipole moment p(in coulomb-meters), or

$$\tau = m \times B$$

for a magnetic dipole moment m(in ampere-square meters).

The resulting torque will tend to align the dipole with the applied field, which in the case of an electric dipole, yields a potential energy of $U = -p \cdot E$.

The energy of a magnetic dipole is similarly $U = -m \cdot B$.

DIPOLE RADIATION

In addition to dipoles in electrostatics, it is also common to consider an electric or magnetic dipole that is oscillating in time. In particular, a harmonically oscillating electric dipole is described by a dipole moment of the form $p = p'(r)e^{-i\omega t}$ where ω is the angular frequency. In vacuum, this produces fields:

$$E = \frac{1}{4\pi\varepsilon_0}\left\{\frac{\omega^2}{c^2 r}\hat{r}\times p\times\hat{r} + \left(\frac{1}{r^3} - \frac{i\omega}{cr^2}\right)\left[3\hat{r}(\hat{r}\cdot p) - p\right]\right\}e^{-iwr/c}$$

$$B = \frac{\omega^2}{4\pi\varepsilon_0 c^3}\hat{r}\times p\left(1 - \frac{c}{iwr}\right)\frac{e^{-iwr/c}}{r}.$$

Far away(for $r\omega/c \gg 1$), the fields approach the limiting form of a radiating spherical wave:

$$B = \frac{\omega^2}{4\pi\varepsilon_0 c^3}(\hat{r}\times p)\frac{e^{iwr/c}}{r}$$

$$E = cB\times\hat{r}$$

which produces a total time-average radiated power P given by

$$P = \frac{\omega^4}{12\pi\varepsilon_0 c^3}|P|^2.$$

This power is not distributed isotropically, but is rather concentrated around the directions lying perpendicular to the dipole moment. Usually such equations are described by spherical harmonics, but they look very different. A circular polarized dipole is described as a superposition of two linear dipoles. A very common source of magnetic field shown in nature is a dipole, with a "South pole" and a "North pole"; terms dating back to the use of magnets as compasses, interacting with the Earth's magnetic field to indicate North and South on the globe.

Since opposite ends of magnets are attracted, the north pole of a magnet is attracted to the south pole of another magnet. Interestingly, this concept of opposite polarities attracting wasn't used in the naming convention for the earth's magnetic field, so the earth's magnetic north pole(in Canada) attracts the magnetic north pole of a compass see North Magnetic Pole. A magnetic field contains energy, and physical systems move toward configurations with lower energy. Therefore, when placed in a magnetic field, a magnetic dipole tends to align itself in opposed polarity to that field, thereby canceling the net field strength as much as possible and lowering the energy stored in that field to a minimum.

For instance, two identical bar magnets placed side-to-side normally line up North to South, resulting in a much smaller net magnetic field, and resist any attempts to reorient them to point in the same direction. The energy required to reorient them in that configuration is then stored in the resulting magnetic field, which is double the strength of the field of each individual magnet.(This is, of course, why a magnet used as a compass interacts with the Earth's magnetic field to indicate North and South). An alternative, equivalent formulation, which is often easier to apply but perhaps offers less insight, is that a magnetic dipole in a magnetic field experiences a torque and a force which can be expressed in terms of the field and the strength of the dipole(i.e., its magnetic dipole moment).

MAGNETIC MONOPOLES

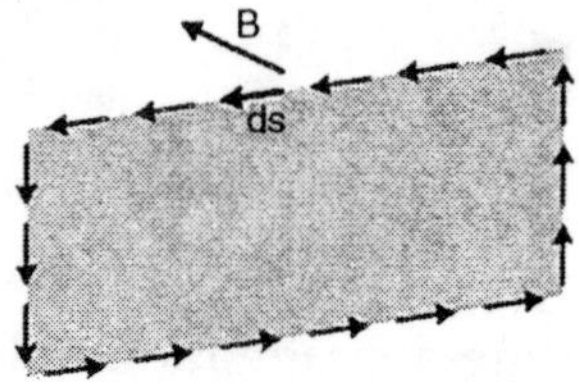

Fig. The Definition of the Circulation, Γ. In this paper, Dirac showed that if magnetic monopoles exist, then that would explain the quantization of electric charge in the universe. Since then, several systematic monopole searches have been performed.

In physics, a magnetic monopole is a hypothetical particle that is a magnet with only one pole.

In more technical terms, it would have a net "magnetic charge". Modern interest in the concept stems from particle theories, notably Grand Unified Theories and superstring theories, which predict their existence.

The classical theory of magnetic charge is as old as Maxwell's equations, but is considered much less important or interesting than the *quantum* theory of magnetic charge, which started with a 1931 paper by Paul Dirac.

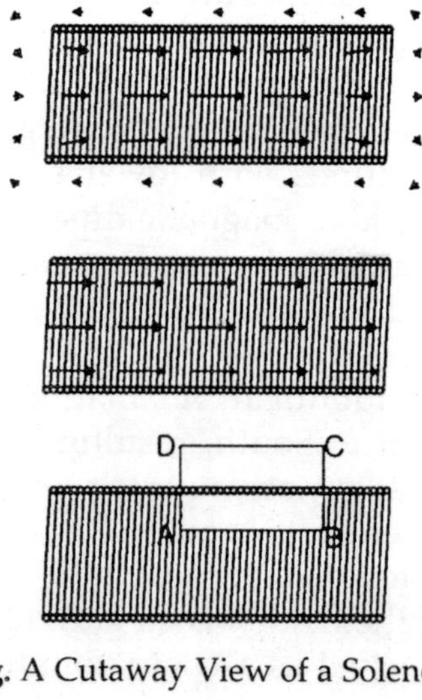

Fig. A Cutaway View of a Solenoid.

Experiments in 1975 and 1982 produced candidate events that were initially interpreted as monopoles, but these are now regarded to be inconclusive.

It therefore remains possible that monopoles do not exist at all. Monopole detection is an open problem in experimental physics. Within theoretical physics, modern approaches agree that monopoles exist, in particular Grand Unified Theories and string theory both require them.

Joseph Polchinski, a prominent string-theorist, described the existence of monopoles as "one of the safest bets that one can make about physics not yet seen". These theories are not necessarily inconsistent with the experimental evidence: in some models magnetic monopoles are unlikely to be observed, because they are too massive to be created in particle accelerators, and too rare in the universe to wander into a particle detector.

Magnets exert forces on one another, similar to electric charges. *Like* poles will repel each other, and *unlike* poles will attract. When a magnet(an object conventionally described as having magnetic north and south poles) is cut in half across the axis joining those "poles", the resulting pieces are two normal(albeit smaller) magnets. Each has its own north pole and south pole. Even atoms have tiny magnetic fields. In the Bohr model of an atom, electrons orbit the nucleus.

The constant change in their motion gives rise to a magnetic field. Permanent magnets have measurable magnetic fields because the atoms and molecules in them are arranged in a way that their individual magnetic fields align, combining to form large aggregate fields.

In this model, the lack of a single pole makes intuitive sense; cutting a bar magnet in half does nothing to the arrangement of the molecules within.

The end result is two magnetic bars whose atoms have the same orientation as before, and therefore generate a magnetic field with the same orientation as the original larger magnet.

MAXWELL'S EQUATIONS

Maxwell's equations of electromagnetism relate the electric and magnetic fields to the motions of electric charges. The standard form of the equations provides for an electric charge, but posits no magnetic charge.

Except for this, the equations are symmetric under interchange of electric and magnetic field. In fact, symmetric equations can be written when all charges are zero, and this is how the wave equation is derived. Fully symmetric equations can also be written if one allows for the possibility of "magnetic charges" analogous to electric charges. If magnetic charges do not exist, or if they exist but where they are not present in a region, then the new variables are zero, and the extended equations reduce to the conventional equations of electromagnetism such as. Classically, the question is *"Why does the magnetic charge always seem to be zero?"*

In SI Units

In SI units, there are two conflicting conventions in use for magnetic charge. In one, magnetic charge has units of webers, while in the other, magnetic charge has units of ampere-meters.

DIRAC'S QUANTIZATION

One of the defining advances in quantum theory was Paul Dirac's work on developing a relativistic quantum electromagnetism. Before his formulation, the presence of electric charge was simply "inserted" into QM, but in 1931 Dirac showed that a discrete charge naturally "falls out" of QM. Consider a system consisting of a single stationary electric monopole(an electron, say) and a single stationary magnetic monopole.

Classically, the electromagnetic field surrounding them has a momentum density given by the Poynting vector, and it also has a total angular momentum, which is proportional to the product $q_e q_m$, and independent of the distance between them.

Quantum mechanics dictates, however, that angular momentum is quantized in units of ', and therefore the product $q_e q_m$ must also be quantized. This means that if even a single magnetic monopole existed in the universe, all electric charges would then be quantized.

What are the units in which magnetic charge would be quantized? Although it would be possible simply to integrate over all space to find the total angular momentum in the above example, Dirac took a different approach, which led to new ideas.

He considered a point-like magnetic charge whose magnetic field behaves as q_m/r and is directed in the radial direction. Because the divergence of B is equal to zero almost everywhere, except for the locus of the magnetic monopole at $r = 0$, one can locally define the vector potential such that the curl of the vector potential A equals the magnetic field B.

However, the vector potential cannot be defined globally precisely because the divergence of the magnetic field is proportional to the delta function at the origin.

We must define one set of functions for the vector potential on the Northern hemisphere, and another set of functions for the Southern hemispheres. These two vector potentials are matched at the equator, and they differ by a gauge transformation. The wave function of an electrically charged particle(a probe) that orbits the equator generally changes by a phase, much like in the Aharonov-Bohm effect.

This phase is proportional to the electric charge q_e of the probe, as well as to the magnetic charges q_m of the source. Dirac was originally considering an electron whose wave function is described by the Dirac equation.

This is known as the Dirac quantization condition. The hypothetical existence of a magnetic monopole would imply that the electric charge must be quantized in certain units; also, the existence of the electric charges implies that the magnetic charges of the hypothetical magnetic monopoles, if they exist, must be quantized in units inversely proportional to the elementary electric charge.

At the time it was not clear if such a thing existed, or even had to. After all, another theory could come along that would explain charge quantization without need for the monopole. The concept remained something of a curiosity. However, in the time since the publication of this seminal work, no other widely accepted explanation of charge quantization has appeared.

If we maximally extend the definition of the vector potential for the Southern hemisphere, it will be defined everywhere except for a semi-infinite line stretched from the origin in the direction towards the Northern pole.

This semi-infinite line is called the Dirac string and its effect on the wave function is analogous to the effect of the solenoid in the Aharonov-Bohm effect. The quantization condition comes from the requirement that the phases around the Dirac string are trivial, which means that the Dirac string must be unphysical.

The Dirac string is merely an artifact of the coordinate chart used and should not be taken seriously. The Dirac monopole is a singular solution of Maxwell's equation(because it requires removing the worldline from spacetime); in more complicated theories, it is superseded by a smooth solution such as the 't Hooft-Polyakov monopole.

TOPOLOGICAL INTERPRETATION

DIRAC STRING

A gauge theory like electromagnetism is defined by a gauge field, which associates a group element to each path in space time. For infinitesimal paths, the group element is close to the identity, while for longer paths the group element is the successive product of the infinitesimal group elements along the way.

In electrodynamics, the group is U(1), unit complex numbers under multiplication. The map from paths to group elements is called the Wilson loop or the holonomy, and for a U(1) gauge group it is the phase factor which the wavefunction of a charged particle acquires as it traverses the path.

So that the phase a charged particle gets when going in a loop is the magnetic flux through the loop. When a small solenoid has a magnetic flux, there are interference fringes for charged particles which go around the solenoid, or around different sides of the solenoid, which reveal its presence.

But if all particle charges are integer multiples of e, solenoids with a flux of $2\pi/e$ have no interference fringes, because the phase factor for any charged particle is. Such a solenoid, if thin enough, is quantum mechanically invisible.

If such a solenoid were to carry a flux of $2\pi/e$, when the flux leaked out from one of its ends it would be indistinguishable from a monopole. Dirac's monopole solution in fact describes an infinitesimal line solenoid ending at a point, and the location of the solenoid is the singular part of the solution, the Dirac string.

Dirac strings link monopoles and antimonopoles of opposite magnetic charge, although in Dirac's version, the string just goes off to infinity. The string is unobservable, so you can put it anywhere, and by using two coordinate patches, the field in each patch can be made nonsingular by sliding the string to where it cannot be seen.

GRAND UNIFIED THEORIES

In a U(1) with quantized charge, the gauge group is a circle of radius $2\pi/e$. Such a U(1) is called compact. Any U(1) which comes from a Grand Unified Theory is compact, because only compact higher gauge groups make sense.

The size of the gauge group is a measure of the inverse coupling constant, so that in the limit of a large volume gauge group, the interaction of any fixed representation goes to zero. The U(1) case is special because all its irreducible representations are the same size— the charge is bigger by an integer amount but the field is still just a complex number— so that in U(1) gauge field theory it is possible to take the decompactified limit with no contradiction.

The quantum of charge becomes small, but each charged particle has a huge number of charge quanta so its charge stays finite. In a non-compact U(1) gauge theory, the charges of particles are generically not integer multiples of a single unit.

Since charge quantization is an experimental certainty, it is clear that the U(1) of electromagnetism is compact. GUTs lead to compact U(1)s, so they explain charge quantization in a way that seems to be logically independent from magnetic monopoles. But the explanation is essentially the same, because in any GUT which breaks down to a U(1) at long distances, there are magnetic monopoles.

The argument is topological:

- The holonomy of a gauge field maps loops to elements of the gauge group. Infinitesimal loops are mapped to group elements infinitesimally close to the identity.
- If you imagine a big sphere in space, you can deform an infinitesimal loop which starts and ends at the north pole as follows: stretch out the loop over the western hemisphere until it becomes a great circle(which still starts and ends at the north pole) then let it shrink back to a little loop while going over the eastern hemisphere. This is called *lassoing the sphere.*
- Lassoing is a sequence of loops, so the holonomy maps it to a sequence of group elements, a continuous path in the gauge group. Since the loop at the beginning of the lassoing is the same as the loop at the end, the path in the group is closed.
- If the group path associated to the lassoing procedure winds around the U(1), the sphere contains magnetic charge. During the lassoing, the holonomy changes by the amount of magnetic flux through the sphere.
- Since the holonomy at the beginning and at the end is the identity, the total magnetic flux is quantized. The magnetic charge is proportional to the number of windings N, the magnetic flux through the sphere is equal to $2\pi N/e$. This is the Dirac quantization condition, and it is a topological condition which demands that the long distance U(1) gauge field configurations are consistent.
- When the U(1) comes from breaking a compact Lie group, the path which winds around the U(1) enough times is topologically trivial in the big group. In a non-U(1) compact lie group, the covering space is a Lie group with the same Lie algebra but where all closed loops are contractible. Lie groups are homogenous, so that any cycle in the group can be moved around so that it starts at the identity, then its lift to the covering group ends at P, which is a lift of the identity. Going around

the loop twice gets you to P, three times to P, all lifts of the identity. But there are only finitely many lifts of the identity, because the lifts can't accumulate. This number of times one has to traverse the loop to make it contractible is small, for example if the GUT group is SO(3), the covering group is SU(2), and going around any loop twice is enough.

- This means that there is a continuous gauge-field configuration in the GUT group allows the U(1) monopole configuration to unwind itself at short distances, at the cost of not staying in the U(1). In order to do this with as little energy as possible, you should only leave the U(1) in the neighborhood of one point, which is called the core of the monopole. Outside the core, the monopole has only magnetic field energy.

So the Dirac monopole is a topological defect in a compact U(1) gauge theory. When there is no GUT, the defect is a singularity— the core shrinks to a point. But when there is some sort of short distance regulator on space time, the monopoles have a finite mass. Monopoles occur in lattice U(1), and there the core size is the lattice size. In general, they are expected to occur whenever there is a short-distance regulator.

STRING THEORY

In our universe, quantum gravity provides the regulator. When gravity is included, the monopole singularity can be a black hole, and for large magnetic charge and mass, the black hole mass is equal to the black hole charge, so that the mass of the magnetic black hole is not infinite. If the black hole can decay completely by Hawking radiation, as required by holography, the lightest charged particles can't be too heavy. The lightest monopole should have a mass less than or comparable to its charge in natural units.

So in a consistent holographic theory, and string theory is the only known example, there are always finite mass monopoles. For ordinary electromagnetism, the mass bound is not very useful because it is about the Planck mass.

MATHEMATICAL FORMULATION

In mathematics, a gauge field is defined as a connection over a principal G-bundle over spacetime. G is the gauge group, and it acts on each fibre of the bundle separately.

A *connection* on a G bundle tells you how to glue F's together at nearby points of M. It starts with a continuous symmetry group G which acts on F, and then it associates a group element with each infinitesimal path. Group multiplication along any path tells you how to move from on point on the bundle to another, by acting the G element of a path on the fibre F.

In mathematics, the definition of bundle is designed to emphasize topology, so the notion of connection is added on as an afterthought. In physics, the connection is the fundamental physical object. Once you have a connection, there are nontrivial bundles which occur as connections of a trivial bundle.

For example, the twisted torus is a connection on a U(1) bundle of a circle on a circle. If space time has no topology, if it is R the space of all possible connections of the *G*-bundle is connected. But consider what happens when we remove a timelike worldline from spacetime. The resulting spacetime is homotopically equivalent to the topological sphere *S*.

A principal *G*-bundle over *S* is defined by covering *S* by two charts, each homeomorphic to the open 2-ball such that their intersection is homeomorphic to the strip $S \times I$. 2-balls are homotopically trivial and the strip is homotopically equivalent to the circle *S*.

So a topological classification of the possible connections is reduced to classifying the transition functions. The transition function maps the strip to G, and the different ways of mapping a strip into G is given by the first homotopy group of *G*. So in the G-bundle formulation, a gauge theory admits Dirac monopoles provided *G* is not simply connected, whenever there are paths that go around the group that cannot be deformed to nothing.

U(1), which has quantized charges, is not simply connected and can have Dirac monopoles while R, its universal covering group, is simply connected, doesn't have quantized charges and does not admit Dirac monopoles.

The mathematical definition is equivalent to the physics definition provided that, following Dirac, gauge fields are allowed which are only defined patch-wise and the gauge field on different patches are glued after a gauge transformation.

This argument for monopoles is a restatement of the lasso argument for a pure U(1) theory. It generalizes to $d + 1$ dimensions with in several ways.

One way is to extend everything into the extra dimensions, so that U(1) monopoles become sheets of dimension d-3. Another way is to examine the type of topological singularity at a point with the homotopy group $\pi_{d-2}(G)$.

GRAND UNIFIED THEORIES

In more recent years, a new class of theories has also suggested the presence of a magnetic monopole. In the early 1970s, the successes of quantum field theory and gauge theory in the development of electroweak and the strong nuclear force led many theorists to move on to attempt to combine them in a single theory known as a grand unified theory, or GUT. Several GUTs were proposed, most of which had the curious feature of suggesting the presence of a real magnetic monopole particle.

More accurately, GUTs predicted a range of particles known as dyons, of which the most basic state is a monopole. The charge on magnetic monopoles predicted by GUTs is either 1 or $2gD$, depending on the theory. The majority of particles appearing in any quantum field theory is unstable, and decays into other particles in a variety of reactions that have to conserve various values.

Stable particles are stable because there are no lighter particles to decay into that still conserve these values. For instance, the electron has a lepton number of 1 and an electric charge of 1, and there are no lighter particles that conserve these values. On the other hand, the muon, essentially a heavy electron, can decay into the electron and is therefore not stable. The dyons in these same theories are also stable, but for an entirely different reason. The dyons are expected to exist as a side effect of the "freezing out" of the conditions of the early universe, or symmetry breaking.

In this model the dyons arise due to the vacuum configuration in a particular area of the universe, according to the original Dirac theory. They remain stable not because of a conservation condition, but because there is no simpler *topological* state to which they can decay. The length scale over which this special vacuum configuration exists is called the *correlation length* of the system.

A correlation length cannot be larger than causality would allow, therefore the correlation length for making magnetic monopoles must be at least as big as the horizon size determined by the metric of the expanding universe. According to that logic, there should be at least one magnetic monopole per horizon volume as it was when the symmetry breaking took place. This leads to a direct prediction of the amount of monopoles in the universe today, which is about 10 times the critical density of our universe. The universe appears to be close to critical density, so monopoles should be fairly common.

For this reason, monopoles became a major interest in the 1970s and 80s, along with the other "approachable" prediction of GUTs, proton decay. The apparent problem with monopoles is resolved by cosmic inflation that greatly reduces the expected abundance of magnetic monopoles. Many of the other particles predicted by these

GUTs were beyond the abilities of current experiments to detect. For instance, a wide class of particles known as the X and Y bosons is predicted to mediate the coupling of the electroweak and strong forces, but these particles are extremely heavy and well beyond the capabilities of any reasonable particle accelerator to create.

MONOPOLE SEARCHES

A number of attempts have been made to detect magnetic monopoles. One of the simplest is to use a loop of superconducting wire that can look for even tiny magnetic sources, a so-called "superconducting quantum interference device", or SQUID.

Given the predicted density, loops small enough to fit on a lab bench would expect to see about one monopole event per year. Although there have been tantalizing events recorded, in particular the event recorded by Blas Cabrera on the night of February 14, 1982(thus, sometimes referred to as the "Valentine's Day Monopole"), there has never been reproducible evidence for the existence of magnetic monopoles. The lack of such events places a limit on the number of monopoles of about 1 monopole per 10 nucleons.

Another experiment in 1975 resulted in the announcement of the detection of a moving magnetic monopole in cosmic rays by the team of Price. Price later retracted his claim, and a possible alternative explanation was offered by Alvarez.

In his paper it was demonstrated that the path of the cosmic ray event that was claimed to be due to a magnetic monopole could be reproduced by a path followed by a Platinum nucleus fragmenting to Osmium and then to Tantalum.

Other experiments rely on the strong coupling of monopoles with photons, as is the case for any electrically charged particle as well. In experiments involving photon exchange in particle accelerators, monopoles should be produced in reasonable numbers, and detected due to their effect on the scattering of the photons.

The probability of a particle being created in such experiments is related to their mass — heavier particles are less likely to be created — so by examining such experiments limits on the mass can be calculated. The most recent such experiments suggest that monopoles with masses below 600 GeV/c^2 do not exist, while upper limits on their mass due to the existence of the universe(which would have collapsed by now if they were too heavy) are about 10 GeV/c^2.

In Popular Culture

- In the turn-based strategy game *Sid Meier's Alpha Centauri,* monopole magnets are one of the researchable technologies. Once a player has developed this technology, that player is able to upgrade roads to magnetic tubes. Units moving along a magnetic tube are able to do so "instantly"(i.e., the movement does not count against the number of moves the unit may be moved each turn).

- In Larry Niven's Known Space monopoles are used widely, specifically in the propulsion systems of slower than light fusion ramjets and various interplanetary craft. They are found and harvested in asteroid belts by belters.
- In the Anime *Outlaw Star* Gene Starwind uses a magnetic monopole to escape from the prison colony of Hecatonchires. Having a pair of monopoles while over the magnetic north of a planet, he discards the south which slams hard into the ground, while the north monopole lifts him and another man up and away.
- In the video game *Star Control II,* magnetic monopoles are a valuable exotic material found on some planets.
- In the online science fiction world-building project Orion's Arm, magnetic monopoles are synthetically created particles that allow for the fabrication of exotic molecules, and are the basis of many advanced technologies.
- In the novel *Omega Minor* by Paul Verhaeghen, one of the key storylines in the book involves a Berlin student's attempts to detect a magnetic monopole, which she eventually manages with the help of a carefully planned nuclear explosion.
- In the video game *Braid,* one of the texts in the epilogue contains magnetic monopoles, naming them among other items desired by the protagonist.

Since a bar magnet gets its ferromagnetism from electrons distributed evenly throughout the bar, when a bar magnet is cut in half, each of the resulting pieces is a smaller bar magnet.

Even though a magnet is said to have a north pole and a south pole, these two poles cannot be separated from each other. A monopole — if such a thing exists — would be a new and fundamentally different kind of magnetic object.

It would act as an isolated north pole, not attached to a south pole, or vice versa. Monopoles would carry "magnetic charge" analogous to electric charge. Despite systematic searches since 1931, as of 2006, they have never been observed, and could very well not exist.

Nevertheless, some theoretical physics models predict the existence of these magnetic monopoles.

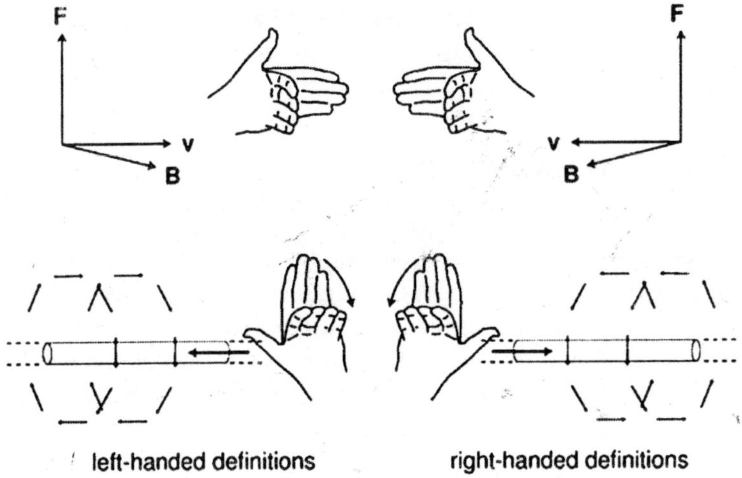

Fig. Left-Handed and Right-Handed Definitions.

Paul Dirac observed in 1931 that, because electricity and magnetism show a certain symmetry, just as quantum theory predicts that individual positive or negative electric charges can be observed without the opposing charge, isolated South or North magnetic poles should be observable. Using quantum theory Dirac showed that if magnetic monopoles exist, then one could explain the quantization of electric charge—that is, why the observed elementary particles carry charges that are multiples of the charge of the electron.

Certain grand unified theories predict the existence of monopoles which, unlike elementary particles, are solitons(localized energy packets). The initial results of using these models to estimate the number of monopoles created in the big bang contradicted cosmological observations – the monopoles would have been so plentiful and massive that they would have long since halted the expansion of the universe.

However, the idea of inflation(for which this problem served as a partial motivation) was successful in solving this problem, creating models in which monopoles existed but were rare enough to be consistent with current observations.

Some organisms can detect magnetic fields, a phenomenon known as magnetoception. Magnetobiology studies magnetic fields as a medical treatment; fields naturally produced by an organism are known as biomagnetism.

The history of electromagnetism, that is the human understanding and recorded use of electromagnetic forces, dates back over two thousand years ago. The ancients must have been acquainted with the effects of atmospheric electricity, as thunderstorms in most southern latitudes are of frequent occurrence, and they also knew of the St. Elmo's fire. They could have had but little knowledge of electricity, and they could not have scientifically explained those phenomena.

ELECTRICITY AND MAGNETISM

Electricity is treated jointly with magnetism, because both generally appear together; wherever the former is in motion, the latter is also present. The phenomenon of magnetism was observed early in the history of magnetism, but was not fully explained until the idea of magnetic induction was developed. The phenomenon of electricity was observed early in the history of electricity, but was not fully explained until the idea of electric charge was fully developed.

Ancient and Classical History

The knowledge of static electricity dates back to the earliest civilizations, but for millennia it remained merely an interesting and mystifying phenomenon, without a theory to explain its behaviour and often confused with magnetism. The ancients were acquainted with other curious properties possessed by two minerals, amber and magnetic iron ore. The former, when rubbed, attracts light bodies: the latter has the power of attracting iron.

Based on his find of an Olmec hematite artifact in Central America, the American astronomer John Carlson has suggested that "the Olmec may have discovered and used the geomagnetic lodestone compass earlier than 1000 BC".

If true, this "predates the Chinese discovery of the geomagnetic lodestone compass by more than a millennium". Carlson speculates that the Olmecs may have used similar artifacts as a directional device for astrological or geomantic purposes, or to orientate their temples, the dwellings of the living or the interments of the dead. The earliest Chinese literature reference to *magnetism* lies in a 4th century BC book called *Book of the Devil Valley Master*: "The lodestone makes iron come or it attracts it."

The discovery of amber and other similar substances in the ancient times suggests the possible perception of it by pre-historic man. The accidental rubbing against the skins with which he clothed himself may have caused an attraction by the resin, thus electrified, of the light fur in sufficiently marked degree to arrest his attention.

Between such a mere observation of the fact, however, and the making of any deduction from it, vast periods may have elapsed; but there came a time at last, when the amber was looked upon as a strange inanimate substance which could influence or even draw to itself other things; and this by its own apparent capacity, and not through any mechanical bond or connection extending from it to them; when it was recognized, in brief, that nature held a lifeless thing showing an attribute of life.

Long before any knowledge of electromagnetism existed, people were indirectly aware of the effects of electricity. Lightning, of course, and certain other manifestations of electricity, were known to the philosophers of ancient times, but to them no thought was more remote than that these manifestations had a common origin.

Ancient Egyptians were aware of shocks when interacting with electric fish(such as the Malapterurus electricus) or other animals(such as electric eels). The shocks from animals were apparent to observers since pre-history by a variety of peoples that came into contact with them. Texts from 2750 BC by the ancient Egyptians, referred to these fish as "thunderer of the Nile", and saw them as the "protectors" of all the other fish.

Possibly the earliest and nearest approach to the discovery of the identity of lightning, and electricity from any other source, is to be attributed to the Arabs, who before the 15th century had the Arabic word for lightning(raad) applied to the Electric ray. According to Thales of Miletus, writing at around 600 BC, noted that a form of electricity was observed by the Ancient Greeks that would cause a particular attraction by rubbing fur on various substances, such as amber. Thales wrote on the effect now known as static electricity. The Greeks noted that the amber buttons could attract light objects such as hair and that if they rubbed the amber for long enough they could even get a spark to jump.

During this time in alchemy and natural philosophy, the existence of a medium of the *æther*, a space-filling substance or field, thought to exist. The electrostatic phenomena was again reported millennia later by Roman and Arabic naturalists and physicians.

Several ancient writers, such as Pliny the Elder and Scribonius Largus, attested to the numbing effect of electric shocks delivered by catfish and torpedo rays. Pliny in his books writes: "The ancient Tuscans by their learning hold that there are nine gods that send forth lightning and those of eleven sorts." This was in general the early pagan idea of lightning.

The ancients held some concept that that shocks could travel along conducting objects. Patients suffering from ailments such as gout or headache were directed to touch electric fish in the hope that the powerful jolt might cure them.

A number of objects found in Iraq in 1938 dated to the early centuries AD(Sassanid Mesopotamia), called the Baghdad Battery, resembles a galvanic cell and is believed by some to have been used for electroplating. The claims are controversial because of supporting evidence and theories for the uses of the artifacts, physical evidence on the objects conducive for electrical functions, and if they were electrical in nature. As a result the nature of these objects is based on speculation, and the function of these artifacts remains in doubt.

Middle Ages and the Renaissance

The attempt to account for magnetic attraction as the working of a soul in the stone led to the first attack of human reason upon superstition and the foundation of philosophy.

After the lapse of centuries, a new capacity of the lodestone became revealed in its polarity, or the appearance of opposite effects at opposite ends; then came the first utilization of the knowledge thus far gained, in the mariner's compass, leading to the discovery of the New World, and the throwing wide of all the portals of the Old to trade and civilization.

In the 11th century, the Chinese scientist Shen Kuo(1031-1095) was the first person to write of the magnetic needle compass and that it improved the accuracy of navigation by employing the astronomical concept of true north(*Dream Pool Essays*, AD 1088), and by the 12th century the Chinese were known to use the lodestone compass for navigation.

In 1187, Alexander Neckham was the first in Europe to describe the compass and its use for navigation. Magnetism was one of the few sciences which progressed in medieval Europe; for in the thirteenth century Peter Peregrinus, a native of Maricourt in Picardy, made a discovery of fundamental importance.

The French 13th century scholar conducted experiments on magnetism and wrote the first extant treatise describing the properties of magnets and pivoting compass needles.

Archbishop Eustathias, of Thessalonica, Greek scholar and writer of the 12th century, records that Woliver, king of the Goths, was able to draw sparks from his body.

The same writer states that a certain philosopher was able while dressing to draw sparks from his clothes, a result seemingly akin to that obtained by Symmer in his silk stocking experiments, a careful account of which may be found in the 'Philosophical Transactions,' 1759.

Italian physician Girolamo Cardano wrote about electricity in *De*

Subtilitate(1550) distinguishing, perhaps for the first time, between electrical and magnetic forces.

Toward the latter part of the 16th century a physician of Queen Elizabeth's time, Dr. William Gilbert, in *De Magnete*, expanded on Cardano's work and coined the New Latin word *electricus* from *$ëåêôñïí(elektron)*, the Greek word for "amber".

The first usage of the word *electricity* is ascribed to Sir Thomas Browne in his 1646 work, *Pseudodoxia Epidemica*. Gilbert undertook a number of careful electrical experiments, in the course of which he discovered that many substances other than amber, such as sulphur, wax, glass, etc., were capable of manifesting electrical properties.

Gilbert also discovered that a heated body lost its electricity and that moisture prevented the electrification of all bodies, due to the now well-known fact that moisture impaired the insulation of such bodies. He also noticed that electrified substances attracted all other substances indiscriminately, whereas a magnet only attracted iron.

Fig. A Generator.

The many discoveries of this nature earned for Gilbert the title of founder of the electrical science. Another pioneers was Robert Boyle, who in 1675 stated that electric attraction and repulsion can act across a vacuum.

One of his important discoveries was that electrified bodies in a vacuum would attract light substances, this indicating that the electrical effect did not depend upon the air as a medium. He also added resin to the then known list of electrics.

This was followed in 1660 by Otto von Guericke, who invented an early electrostatic generator. By the end of the 17th Century, researchers had developed practical means of generating electricity by friction in electrostatic generator, but the development of electrostatic machines did not begin in earnest until the 18th century, when they became fundamental instruments in the studies about the new science of electricity.

Electrostatic generators operate by using manual(or other) power to transform mechanical work into electric energy. They develop electrostatic charges of opposite signs rendered to two conductors, using only electric forces.

18TH CENTURY

Early 1700s

Isaac Newton contended that light was made up of numerous small particles. This could explain such features as light's ability to travel in straight lines and reflect off surfaces.

This theory was known to have its problems: although it explained reflection well, its explanation of refraction and diffraction was less satisfactory. In order to explain refraction, Newton's *Opticks*(1704) postulated an "Aethereal Medium" transmitting vibrations *faster* than light, by which light, when overtaken, is put into "Fits of easy Reflexion and easy Transmission", which caused refraction and diffraction.

IMPROVING THE ELECTRIC MACHINE

The electric machine was subsequently improved by Hawkesbee or Haukesbee, Litzendorf, and by Prof. George Mathias Boze, about 1750. Litzendorf substituted a glass ball for the sulphur ball of Guericke. Boze was the first to employ the "prime conductor" in such machines, this consisting of an iron rod held in the hand of a person whose body was insulated by standing on a cake of resin.

Dr. Ingenhousz, in 1746, invented electric machines made of plate glass. Experiments with the electric machine were largely aided by the discovery of the property of a glass plate, when coated on both sides with tinfoil, of accumulating a charge of electricity when connected with a source of electromotive force.

The electric machine was soon further improved by Andrew Gordon, a Scotchman, Professor at Erfurt, who substituted a glass cylinder in place of a glass globe; and by Giessing of Leipzig who added a "rubber" consisting of a cushion of woolen material. The collector, consisting of a series of metal points, was added to the machine by Benjamin Wilson about 1746, and in 1762, John Canton of England(also the inventor of the first pith-ball electroscope) improved the efficiency of electric machines by sprinkling an amalgam of tin over the surface of the rubber.

Electrics and Non-Electrics

In 1729, Stephen Gray conducted a series of experiments that demonstrated the difference between conductors and non-conductors(insulators), showing amongst other things that a metal wire and even pack thread conducted electricity, whereas silk did not. In one of his experiments he sent an electric current through 800 feet of hempen thread which was suspended at intervals by loops of silk thread. When he tried to conduct the same experiment substituting the silk for finely spun brass wire, he found that the electrical current was no longer carried throughout the hemp cord, but instead seemed to vanish into the brass wire. From this experiment he classified substances into two categories: "electrics" like glass, resin and silk and "non-electrics" like metal and water. "Electrics" conducted charges while "non-electrics" held the charge.

Vitreous and Resinous

Intrigued by Gray's results, in 1732, C. F. du Fay began to conduct several experiments. In his first experiment, Du Fay concluded that all objects except metals, animals, and liquids could be electrified by rubbing and that metals, animals and liquids could be electrified by means of an electric machine, thus discrediting Gray's "electrics" and "non-electrics" classification of substances.

In 1737 Du Fay and Hawksbee independently discovered what they believed to be two kinds of frictional electricity; one generated from rubbing glass, the other from rubbing resin.

From this, Du Fay theorized that electricity consists of two electrical fluids, "vitreous" and "resinous", that are separated by friction and that neutralize each other when combined. This two-fluid theory would later give rise to the concept of *positive* and *negative* electrical charges devised by Benjamin Franklin.

Leyden Jar

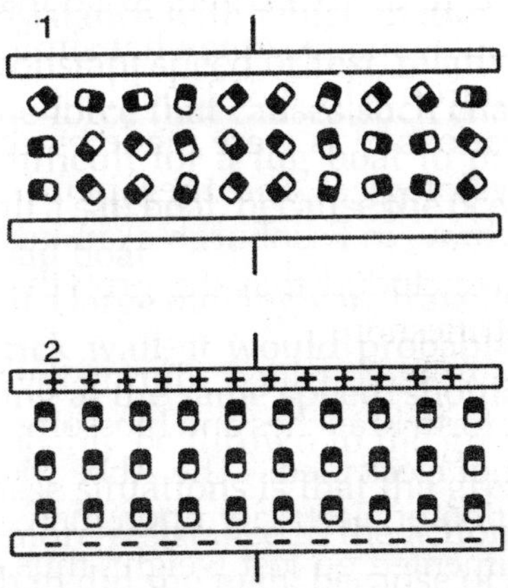

Fig. A Capacitor with a Dielectric Between the Plates.

The Leyden jar, a type of capacitor for electrical energy in large quantities, was invented at Leiden University by Pieter van Musschenbroek in 1745. William Watson, when experimenting with the Leyden jar, discovered in 1747 that a discharge of static electricity was equivalent to an electric current.

The capacitive property, now and for many years availed of in the electric condenser, was first observed by Von Kleist of Leyden in 1754. Von Kleist happened to hold, near his electric machine, a small bottle, in the neck of which there was an iron nail.

Touching the iron nail accidentally with his other hand he received a severe electric shock. In much the same way Prof. Pieter van Musschenbroeck assisted by Cunaens received a more severe shock from a somewhat similar glass bottle. Sir William Watson of England greatly improved this device, by covering the bottle, or jar, outside and in with tinfoil. This piece of electrical apparatus will be easily recognized as the well-known Leyden jar, so called by the Abbot Nollet of Paris, after the place of its discovery.

In 1741, Ellicott "proposed to measure the strength of electrification by its power to raise a weight in one scale of a balance while the other was held over the electrified body and pulled to it by its attractive power". The Sir William Watson already mentioned conducted numerous experiments, about 1749, to ascertain the velocity of electricity in a wire, which experiments, although perhaps not so intended, also demonstrated the possibility of transmitting signals to a distance by electricity.

In these experiments an insulated wire 12,276 feet in length was employed and the transmission of a signal from one end of the wire to the other appeared to the observers to be instantaneous. Monnicr in France had previously made somewhat similar experiments, sending shocks through an iron wire 1,319 feet long.

About 1750 various tests were made by different experimenters to ascertain the physiological and therapeutical effects of electricity. Mainbray(or Mowbray) in Edinburgh examined the effects of electricity upon plants and concluded that the growth of two myrtle trees was quickened by electrification.

These myrtles were electrified "during the whole month of October, 1746, and they put forth branches and blossoms sooner than other shrubs of the same kind not electrified." The Abbé Menon tried the effects of a continued application of electricity upon men and birds and found that the subjects experimented on lost weight, thus apparently showing that electricity quickened the excretions.

The efficacy of electric shocks in cases of paralysis was tested in the county hospital at Shrewsbury, England, with rather poor success. In one case reported a palsied arm was somewhat improved, but the dread of the shocks became so great that the patient preferred to forego a possible cure rather than undergo further treatment.

In another case of partial paralysis the electric treatment was followed by temporary total paralysis. A second application of this treatment was again followed by total paralysis, whereupon the further use of electricity in this case was stopped. For further accounts of the early use of electricity as a remedial agent the reader may consult De la Rive's 'Electricity.'

Late 1700s

In 1752, Benjamin Franklin is frequently confused as the key luminary behind electricity. William Watson and Benjamin Franklin share the discovery of electrical potentials. Benjamin Franklin promoted his investigations of electricity and theories through the famous, though extremely dangerous, experiment of flying a kite through a storm-threatened sky.

A key attached to the kite string sparked and charged a Leyden jar, thus establishing the link between lightning and electricity. Following these experiments he invented a lightning rod. It is either Franklin(more frequently) or Ebenezer Kinnersley of Philadelphia(less frequently) who is considered as the establisher of the convention of positive and negative electricity.

Theories regarding the nature of electricity were quite vague at this period, and those prevalent were more or less conflicting. Franklin considered that electricity was an imponderable fluid pervading everything, and which, in its normal condition, was uniformly distributed in all substances.

He assumed that the electrical manifestations obtained by rubbing glass were due to the production of an excess of the electric fluid in that substance and that the manifestations produced by rubbing wax were due to a deficit of the fluid.

This theory was opposed by the "two-fluid" theory due to Robert Symmer, 1759. By Symmer's theory the vitreous and resinous electricities were regarded as imponderable fluids, each fluid being composed of mutually repellent particles while the particles of the opposite electricities arc mutually attractive.

When the two fluids unite by reason of their attraction for one another, their effect upon external objects is neutralized. The act of rubbing a body decomposes the fluids one of which remains in excess on the body and manifests itself as vitreous or resinous electricity.

Up to the time of Franklin's historic kite experiment the identity of the electricity developed by rubbing and by electric machines(frictional electricity), with lightning had not been generally established. Dr. Wall, Abbot Nollet, Hawkesbee, Gray and Winckler had indeed suggested the resemblance between the phenomena of "electricity" and "lightning," Gray having intimated that they only differed in degree.

It was doubtless Franklin, however, who first proposed tests to determine the sameness of the phenomena. In a letter to Peter Comlinson, London, 19 Oct. 1752.

Franklin, referring to his kite experiment, wrote, "At this key the phial(Leyden jar) may be charged; and from the electric fire thus obtained spirits may be kindled, and all the other electric experiments be formed which are usually done by the help of a rubbed glass globe or tube, and thereby the sameness of the electric matter with that of lightning be completely demonstrated."

Dalibard, at Marley, near Paris, on 10 May 1742, by means of a vertical iron rod 40 feet long, obtained results corresponding to those recorded by Franklin and somewhat prior to the date of Franklin's experiment. Franklin's important demonstration of the sameness of frictional electricity and lightning doubtless added zest to the efforts of the many experimenters in this field in the last half of the 18th century, to advance 'the progress of the science.

Franklin's observations aided later scientists such as Michael Faraday, Luigi Galvani, Alessandro Volta, André-Marie Ampère, and Georg Simon Ohm whose work provided the basis for modern electrical technology. The work of Faraday, Volta, Ampere, and Ohm is honoured by society, in that fundamental units of electrical measurement are named after them. Others would also advance the field of knowledge including those workers Watson, Boze, Smeaton, Le Monnicr, De Romas, Jallabert, Beccaria, Cavallo, John Canton, Robert Symmer, Nollet, Winckler, Richman, Dr. Wilson, Kinnersley, Priestley, Aepinus, Délavai, Cavendish, Coulomb, Volta and Galvani.

A description of many of the experiments and discoveries of these early workers in the fields of electrical science and art will be found in the scientific publications of the time; notably the 'Philosophical Transactions,1 Philosophical Magazine, Cambridge Mathematical Journal, Young's 'Natural Philosophy,' Priestley's 'History of Electricity,' ' Franklin's 'Experiments and Observations on Electricity,' Cavalli's 'Treatise on Electricity,' De la Rive's 'Treatise on Electricity.' Among the more important of the electrical experiments and researches at this period were those of Francis Aepinus, a noted German scholar(1724-1802) and Henry Cavendish of London, England.

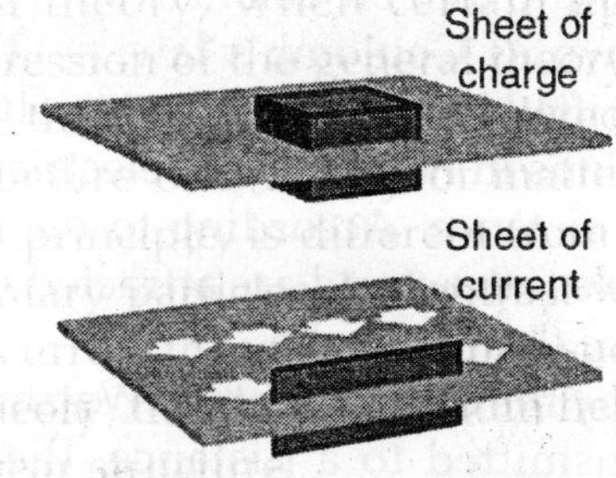

Fig. A Gaussian Surface and an Ampèrian Surface.

To Aepinus is accorded the credit of having been the first to conceive the view of the reciprocal relationship of electricity and magnetism. In his work 'Tentamen Theoria Electricitatis et Magnetism!,' published in Saint Petersburg, 1759.

He gives the following amplification of Franklin's theory, which in some of its features is measurably in accord with present day views : "The particles of the electric fluid repel each other and attract and are attracted by the particles of all bodies with a force that decreases in proportion as the distance increases; the electric fluid exists in the pores of bodies; it moves unobstructedly through non-electric(conductors), but moves with difficulty in insulators; the manifestations of electricity are due to the unequal distribution of the fluid in a body, or to the approach of bodies unequally charged with the fluid."

Aepinus formulated a corresponding theory of magnetism excepting that in the case of magnetic phenomena the fluids only acted on the particles of iron. He also made numerous electrical experimens, amongst others those apparently showing that in order to manifest electrical effects tourmalin requires to be heated to a temperature between 37.5° and 100° C. In fact, tourmalin remains unelectrified when its temperature is uniform, but manifests electrical properties when its temperature is rising or falling. Crystals which manifest electrical properties in this way are termed pyro-electrics, amongst which, besides tourmalin, are sulphate of quinine and quartz.

Cavendish independently conceived a theory of electricity nearly akin to that of Aepinus.

He also(1784) was perhaps the first to utilize the electric spark to produce the explosion of hydrogen and oxygen in the proper proportions to produce pure water. The same philosopher also discovered the inductive capacity of dielectrics(insulators) and as early as 1778 measured the specific inductive capacity for beeswax and other substances by comparison with an air condenser.

About 1784 C. A. Coulomb, after whom is named the electrical unit of quantity, devised the torsion balance, by means of which he discovered what is known as Coulomb's law; — The force exerted between two small electrified bodies varies inversely as the square of the distance; not as Aepinus in his theory of electricity had assumed, merely inversely as the distance. According to the theory advanced by Cavendish "the particles attract and are attracted inversely as some less power of the distance than the cube."

With the discovery, by the experiments of Watson and others, that electricity could be transmitted to a distance, the idea of making practical use of this phenomenon began, about 1753, to engross the minds of "inquisitive" persons, and to this end suggestions looking to the employment of electricity in the transmission of intelligence were made. The first of the methods devised for this purpose was probably that, due to besage(1774).

This method consisted in the employment of 24 wires, insulated from one another and each of which had a pith ball connected to its distant end. Each wire represented a letter of the alphabet. To send a message, a desired wire was charged momentarily with electricity from an electric machine, whereupon the pith ball connected to that wire would fly out; and in this way messages were transmitted. Other methods of telegraphing in which frictional electricity was employed were also tried, some of which are described in the article on the telegraph.

Hitherto the only electricity known was that developed by friction or rubbing, which was therefore termed frictional electricity. We now come to the era of galvanic or voltaic electricity. Volta discovered that chemical reactions could be used to create positively charged anodes and negatively charged cathodes. When a conductor was attached between these, the difference in the electrical potential(also known as voltage) drove a current between them through the conductor. The potential difference between two points is measured in units of volts in recognition of Volta's work.

The first mention of voltaic electricity, although not recognized as such at the time, was probably made by Sulzer in 1767, who on placing a small disc of zinc under his tongue and a small disc of copper over it, observed a peculiar taste when the respective metals touched at their edges. Sulzer assumed that when the metals came together they were set into vibration, this acting upon the nerves of the tongue, producing the effects noticed. In 1790 Prof. Luigi Alyisio Galvani of Bologna on one occasion, while conducting experiments on "animal electricity," as he termed it, to which his attention had been turned by the twitching of a frog's legs in the presence of an electric machine, observed that the muscles of a frog which was suspended on an iron balustrade by a copper hook that passed through its dorsal column underwent lively convulsions without any extraneous cause; the electric machine being at this time absent.

To account for this phenomenon Galvani assumed that electricity of opposite kinds existed in the nerves and muscles of the frog; the muscles and nerves constituting the charged coatings of a Leyden jar. Galvani published the results of his discoveries, together with his hypothesis, which at once engrossed the attention of the physicists of that time; the most prominent of whom, Alexander Volta, professor of physics at Pavia, contended that the results observed by Galvani were due to the two metals, copper and iron, acting as "electromotors," and that the muscles of the frog played the part of a conductor, completing the circuit. This precipitated a long discussion between the adherents of the conflicting views; one set of adherents holding with Volta that the electric current was the result of an electromotive force of contact at the two metals; the other set adopting a modification of Galvani's view and asserting that the current was due to a chemical affinity between the metals and the acids in the pile.

Michael Faraday wrote in the preface to his "Experimental Researches", relative to the question whether metallic contact is or is not productive of a part of the electricity of the voltaic pile: I see no reason as yet to alter the opinion I have given;... but the point itself is of such great importance that I intend at the first opportunity renewing the inquiry, and, if I can, rendering the proofs either on the one side or the other, undeniable to all." Even Faraday himself, however, did not settle the controversy, and while the views of the advocates on both sides of the question have undergone modifications, as subsequent investigations and discoveries demanded, up to the present day diversity of opinion on these points continues to crop out.

Volta made numerous experiments in support of his theory and ultimately developed the pile or battery, which was the precursor of all subsequent chemical batteries, and possessed the distinguishing merit of being the first means by which a prolonged continuous current of electricity was obtainable. Volta communicated a description of his pile to the Royal Society of London and shortly thereafter Nicholson and Cavendish(1780) produced the decomposition of water by means of the electric current, using Volta's pile as the source of electromotive force.

19TH CENTURY

Early 1800s

In 1800 Alessandro Volta constructed the first device to produce a large electric current, later known as the electric battery. Napoleon, informed of his works, summoned him in 1801 for a command performance of his experiments. He received many medals and declorations, including the Légion d'honneur.

Davy in 1806, employing a voltaic pile of approximately 250 cells, or couples, decomposed potash and soda, showing that these substances were respectively the oxides of potassium and sodium, which metals previously had been unknown.

These experiments were the beginning of electrochemistry, the investigation of which Faraday took up, and concerning which in 1833 he announced his important law of electrochemical equivalents, viz.: "The same quantity of electricity — that is, the same electric current — decomposes chemically equivalent quantities of all the bodies which it traverses; hence the weights of elements separated in these electrolytes are to each other as their chemical equivalents." Employing a battery of 2,000 elements of a voltaic pife Humphrey Davy in 1809 gave the first public demonstration of the electric arc light, using for the purpose charcoal enclosed in a vacuum.

Somewhat singular to note, it was not until many years after the discovery of the voltaic pile that the sameness of annual and frictional electricity with voltaic electricity was clearly recognized and demonstrated.

Thus as late as January 1833 we find Faraday writing in a paper on the electricity of the electric ray. "After an examination of the experiments of Walsh, Ingenhousz, Henry Cavendish, Sir H. Davy, and Dr. Davy, no doubt remains on my mind as to the identity of the electricity of the torpedo with common(frictional) and voltaic electricity; and I presume that so little will remain on the mind of others as to justify my refraining from entering at length into the philosophical proof of that identity. The doubts raised by Sir Humphrey Davy have been removed by his brother, Dr. Davy; the results of the latter being the reverse of those of the former.... The general conclusion which must, I think, be drawn from this collection of facts(a table showing the similarity, of properties of the diversely named electricities) is, that electricity, whatever may be its source, is identical in its nature."

It is proper to state, however, that prior to Faraday's time the similarity of electricity derived from different sources was more than suspected. Thus, William Hyde Wollaston, wrote in 1801: "This similarity in the means by which both electricity and galvanism(voltaic electricity) appear to be excited in addition to the resemblance that has been traced between their effects shows that they are both essentially the same and confirm an opinion that has already been advanced by others, that all the differences discoverable in the effects of the latter may be owing to its being less intense, but produced in much larger quantity." In the same paper Wollaston describes certain experiments in which he uses very fine wire in a solution of sulphate of copper through which he passed electric currents from an electric machine. This is interesting in connection with the later day use of almost similarly arranged fine wires in electrolytic receivers in wireless, or radio-telegraphy.

In the first half of the 19th century many very important additions were made to the world's knowledge concerning electricity and magnetism. For example, in 1819 Hans Christian Oersted of Copenhagen discovered the deflecting effect of an electric current traversing a wire upon- a suspended magnetic needle.

This discovery gave a clue to the subsequently proved intimate relationship between electricity and magnetism which was promptly followed up by Ampère who shortly thereafter(1821) announced his celebrated theory of electrodynamics, relating to the force that one current exerts upon another, by its electro-magnetic effects, namely:

- Two parallel portions of a circuit attract one another if the currents in them are flouring in the same direction, and repel one another if the currents flow in the opposite direction.
- Two portions of circuits crossing one another obliquely attract one another if both the currents flow either towards or from the point of crossing, and repel one another if one flows to and the other from that point.
- When an element of a circuit exerts a force on another element of a circuit, that force always tends to urge the latter in a direction at right angles to its own direction.

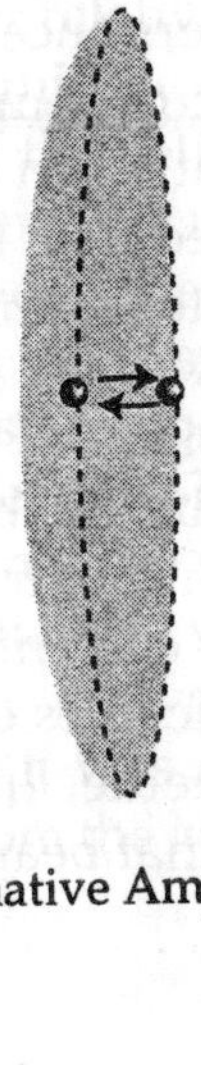

Fig. An Alternative Ampèrian Surface.

Professor Seebeck, of Berlin, in 1821 discovered that when heat is applied to the junction of two metals that had been soldered together an electric current is set up.

This is termed Thermo-Electricity. Seebeck's device consists of a strip of copper bent at each end and soldered to a plate of bismuth. A magnetic needle is placed parallel with the copper strip. When the heat of a lamp is applied to the junction of the copper and bismuth an electric current is set up which deflects the needle.

Peltier in 1834 discovered an effect opposite to the foregoing, namely, that when a current is passed through a couple of dissimilar metals the temperature is lowered or raised at the junction of the metals, depending on the direction of the current.

This is termed the Peltier "effect". The variations of temperature arc found to be proportional to the strength of the current and not to the square of the strength of the current as in the case of heat due to the ordinary resistance of a conductor.

This latter is the C2R law, discovered experimentally in 1841 by the English physicist, Joule. In other words, this important law is that the heat generated in any part of an electric circuit is directly proportional to the product of the resistance of this part of the circuit and to the square of the strength of current flowing in the circuit. In 1822 Sweiprger devised the first galvanometer.

This instrument was subsequently much improved by Wilhelm Weber(1833). In 1825 William Sturgeon of Woolwich, England, invented the horseshoe and straight bar electromagnet, receiving therefor the silver medal of the Society of Arts. In 1837 Gauss and Weber(both noted workers of this period) jointly invented a reflecting galvanometer for telegraph purposes.

This was the forerunner of the Thomson reflecting and other exceedingly sensitive galvanometers once used in submarine signaling and still widely employed in electrical measurements. Arago in 1824 made the important discovery that when a copper disc is rotated in its own plane, and if a magnetic needle be freely suspended on a pivot over the disc, the needle will rotate with the disc. If on the other hand the needle is fixed it will tend to retard the motion of the disc. This effect was termed Arago's rotations.

Futile attempts were made by Babbage, Barlow, Herschel and others to explain this phenomenon. The true explanation was reserved for Faraday, namely, that electric currents are induced in the copper disc by the cutting of the magnetic lines of force of the needle, which currents in turn react on the needle. In 1827 George Simon Ohm announced the now famous law that bears his name, that is:

Electromotive force = Current × Resistance

FARADAY AND HENRY

In 1831 began the epoch-making researches of Michael Faraday, the famous pupil and successor of Humphrey Davy at the head of the Royal Institution, London, relating to electric and electromagnetic induction. Faraday's studies and researches extended from 1831 to 1855 and a detailed description of his experiments, deductions and speculations arc to be found in his compiled papers, entitled Experimental Researches in Electricity.'

Faraday was by profession a chemist. He was not in the remotest degree a mathematician in the ordinary sense — indeed it is a quest on if in all his writings there is a single mathematical formula.

The experiment which led Faraday to the discovery of electric induction was made as follows: He constructed what is now and was then termed an induction coil, the primary and secondary wires of which were wound on a wooden bobbin, side by side, and insulated from one another. In the circuit of the primary wire he placed a battery of approximately 100 cells.

In the secondary wire he inserted a galvanometer. On making his first test he observed no results, the galvanometer remaining quiescent, but on increasing the length of the wires he noticed a deflection of the galvanometer in the secondary wire when the circuit of the primary wire was made and broken. This was the first observed instance of the development of electromotive force by electromagnetic induction.

He also discovered that induced currents are established in a second closed circuit when the current strength is varied in the first "wire, and that the direction of the current in the secondary circuit is opposite to that in the first circuit. Also that a current is induced in a secondary circuit when another circuit carrying a current is moved to and from the first circuit, and that the approach or withdrawal of a magnet to or from a closed circuit induces momentary currents in the latter.

In short, within the space of a few months Faraday discovered by experiment virtually all the laws and facts now known concerning electro-magnetic induction and magneto-electric induction. Upon these discoveries, with scarcely an exception, depends the operation of the telephone, the dynamo machine, and incidental to the dynamo electric machine practically all the gigantic electrical industries of the world, including electric lighting, electric traction, the operation of electric motors for power purposes, and electro-plating, electrotyping, etc.

In his investigations of the peculiar manner in which iron filings arrange themselves on a cardboard or glass in proximity to the poles of a magnet, Faraday conceived the idea of magnetic "lines of force" extending from pole to pole of the magnet and along which the filings tend to place themselves.

On the discovery being made that magnetic effects accompany the passage of an electric current in a wire, it was also assumed that similar magnetic lines of force whirled around the wire. For convenience and to account for induced electricity it was then assumed that when these lines of force are «cut" by a wire in passing across them or when the lines of force in rising and falling cut the wire, a current of electricity is developed, or to be more exact, an electromotive force is developed in the wire that sets up a current in a closed circuit.

Faraday advanced what has been termed the molecular theory of electricity which assumes that electricity is the manifestation of a peculiar condition of the molecule of the body rubbed or the ether surrounding the body.

Faraday also, by experiment, discovered paramagnetism and diamagnetism, namely, that all solids and liquids are either attracted or repelled by a magnet. For example, iron, nickel, cobalt, manganese, chromium, etc., are paramagnetic(attracted by magnetism), whilst other substances, such as bismuth, phosphorus, antimony, zinc, etc., are repelled by magnetism or are diamagnetic.

Brugans of Leyden in 1778 and Le Baillif and Becquerel in 1827 had previously discovered diamagnetism in the case of bismuth and antimony. Faraday also rediscovered specific inductive capacity in 1837, the results of the experiments by Cavendish not having been published at that time.

He also predicted the retardation of signals on long submarine cables due to the inductive effect of the insulation of the cable, in other words, the static capacity of the cable.

The 25 years immediately following Faraday's discoveries of electric induction were fruitful in the promulgation of laws and facts relating to induced currents and to magnetism. In 1834 Lenz and Jacobi independently demonstrated the now familiar fact that the currents induced in a coil are proportional to the number of turns in the coil. Lenz also announced at that time the important law that, in all cases of electromagnetic induction the induced currents have such a direction that their reaction tends to stop the motion that produces, them, a law that was perhaps deducible from Faraday's explanation of Arago's rotations.

In 1845 Joseph Henry, the American physicist, published an account of his valuable and interesting experiments with induced currents of a high order, showing that currents could be induced from the secondary of an induction coil to the primary of a second coil, thence to its secondary wire, and so on to the primary of a third coil, etc.

The electromagnetic theory of light adds to the old undulatory theory an enormous province of transcendent interest and importance; it demands of us not merely an explanation of all the phenomena of light and radiant heat by transverse vibrations of an elastic solid called ether, but also the inclusion of electric currents, of the permanent magnetism of steel and lodestone, of magnetic force, and of electrostatic force, in a comprehensive ethereal dynamics.

MIDDLE 1800S

Up to the middle of the 19th century, indeed up to about 1870, electrical science was, it may be said, a sealed book to the majority of electrical workers. Prior to this time a number of handbooks had been published on electricity and magnetism, notably Aug. de La Rive's exhaustive 'Treatise on Electricity,' 1851 and(in the French) 1835; Beer's Einleitung in die Electrostatik, Wiedemann's 'Galvanismus,' and Reiss' 'Reibungsal-elektricitat.'

But these works consisted in the main in details of experiments with electricity and magnetism, and but little with the laws and facts of those phenomena. Abria published the results of some researches into the laws of induced currents, but owing to their complexity of the investigation it was not productive of very notable results. Around the mid-1800s, Fleeming Jenkin's work on 'Electricity and Magnetism' and Clerk Maxwell's 'Treatise on Electricity and Magnetism' were published.

These books were departures from the beaten path. As Jenkin states in the preface to his work the science of the schools was so dissimilar from that of the practical electrician that it was quite impossible to give students sufficient, or even approximately sufficient, textbooks.

A student he said might have mastered De la Rive's large and valuable treatise and yet feel as if in an unknown country and listening to an unknown tongue in the company of practical men.

As another writer has said, with the coming of Jenkin's and Maxwell's books all impediments in the way of electrical students were removed, "the full meaning of Ohm's law becomes clear; electromotive force, difference of potential, resistance, current, capacity, lines of force, magnetization and chemical affinity were measurable, and could be reasoned about, and calculations could be made about them with as much certainty as calculations in dynamics".

About 1850 Kirchoff published his laws relating to branched or divided circuits. He also showed mathematically that according to the then prevailing electrodynamic theory, electricity would be propagated along a perfectly conducting wire with the velocity of light.

Helmholtz investigated mathematically the effects of induction upon the strength of current and deduced therefrom equations, which experiment confirmed, showing amongst other important points the retarding effect of self-induction under certain conditions of the circuit. In 1853 Sir William Thomson(later Lord Kelvin) predicted as a result of mathematical calculations the oscillatory nature of the electric discharge of a condenser circuit. To Henry, however, belongs the credit of discerning as a result of his experiments in 1842 the oscillatory nature of the Leyden jar discharge.

He wrote: The phenomena require us to admit the existence of a principal discharge in one direction, and then several reflex actions backward and forward, each more feeble than the preceding, until the equilibrium is obtained.

These oscillations were subsequently observed by Fcddersen(1857) who using a rotating concave mirror projected an image of the electric spark upon a sensitive plate, thereby obtaining a photograph of the spark which plainly indicated the alternating nature of the discharge. Sir William Thomson was also the discoverer of the electric convection of heat(the "Thomson" effect). He designed for electrical measurements of precision his quadrant and absolute electrometers. The reflecting galvanometer and siphon recorder, as applied to submarine cable signaling, are also due to him.

About 1876 Prof. H. A. Rowland of Baltimore demonstrated the important fact that a static charge carried around produces the same magnetic effects as an electric current. The Importance of this discovery consists in that it may afford a plausible theory of magnetism, namely, that magnetism may be the result of directed motion of rows of molecules carrying static charges.

After Faraday's discovery that electric currents could be developed in a wire by causing it to cut across the lines of force of a magnet, it was to be expected that attempts would be made to construct machines to avail of this fact in the development of voltaic currents.

The first machine of this kind was due to Pixii, 1832. It consisted of two bobbins of iron wire, opposite which the poles of a horseshoe magnet were caused to rotate.

As this produced in the coils of the wire an alternating current, Pixii arranged a commutating device(commutator) that converted the alternating current of the coils or armature into a direct current in the external circuit. This machine was followed by improved forms of magneto-electric machines due to Ritchie, Saxton, Clarke, Stohrer 1843, Nollet 1849, Shepperd 1856, Van Maldern, Siemens, Wilde and others. A notable advance in the art of dynamo construction was made by Mr. S. A. Varley in 1866 and by Dr. Charles William Siemens and Mr. Charles Wheatstone, who independently discovered that when a coil of wire, or armature, of the dynamo machine is rotated between the poles(or in the "field") of an electromagnet, a weak current is set up in the coil due to residual magnetism in the iron of the electromagnet, and that if the circuit of the armature be connected with the circuit of the electromagnet, the weak current developed in the armature increases the magnetism in the field.

This further increases the magnetic lines of force in which the armature rotates, which still further increases the current in the electromagnet, thereby producing a corresponding increase in the field magnetism, and so on, until the maximum electromotive force which the machine is capable of developing is reached.

By means of this principle the dynamo machine develops its own magnetic field, thereby much increasing its efficiency and economical operation. Not by any means, however, was the dynamo electric machine perfected at the time mentioned.

In 1860 an important improvement had been made by Dr. Antònio Pacinotti of Pisa who devised the first electric machine with a ring armature. This machine was first used as an electric motor, but afterward as a generator of electricity.

The discovery of the principle of the reversibility of the dynamo electric machine(variously attributed to Walenn 1860; Pacinotti 1864 ; Fontaine, Gramme 1873; Deprez 1881, and others) whereby it may be used as an electric motor or as a generator of electricity has been termed one of the greatest discoveries of the 19th century.

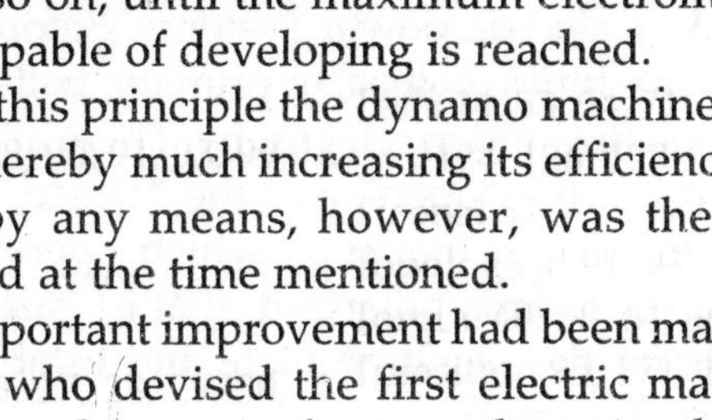

Fig. A Transformer.

In 1872 the drum armature was devised by Heffner-Altneck. This machine in a modified form was subsequently known as the Siemens dynamo. These machines were presently followed by the Schuckert, Gulcher, Fein, Brush, Hochhausen, Edison and the dynamo machines of numerous other inventors. In the early days of dynamo machine construction the machines were mainly arranged as direct current generators, and perhaps the most important application of such machines at that time was in electro-plating, for which purpose machines of low voltage and large current strength were employed.

Beginning about 1887 alternating current generators came into extensive operation and the commercial development of the transformer, by means of which currents of low voltage and high current strength are transformed to currents of high voltage and low current strength, and vice-versa, in time revolutionized the transmission of electric power to long distances. Likewise the introduction of the rotary converter(in connection with the "step-down" transformer) which converts alternating currents into direct currents(and vice-versa) has effected large economies in the operation of electric power systems. Before the introduction of dynamo electric machines, voltaic, or primary, batteries were extensively used for electro-plating and in telegraphy. There are two distinct types of voltaic cells, namely, the "open" and the "closed," or "constant," type. The open type in brief is that type which operated on closed circuit becomes, after a short time, polarized; that is, gases are liberated in the cell which settle on the negative plate and establish a resistance that reduces the current strength.

After a brief interval of open circuit these gases are eliminated or absorbed and the cell is again ready for operation. Closed circuit cells are those in which the gases in the cells are absorbed as quickly as liberated and hence the output of the cell is practically uniform. The Leclanché and Daniell cells, respectively, are familiar examples of the "open" and "closed" type of voltaic cell. The "open" cells are used very extensively at present, especially in the dry cell form, and in annunciator and other open circuit signal systems. Batteries of the Daniell or "gravity" type were employed almost generally in the United States and Canada as the source of electromotive force in telegraphy before the dynamo machine became available, and still are largely used for this service or as "local" cells. Batteries of the "gravity" and the Edison-Lalande types are still much used in "closed circuit" systems. In the late 19th century, the term luminiferous aether, meaning light-bearing aether, was the term used to describe a medium for the propagation of light. The word *aether* stems via Latin from the Greek áéèÞñ, from a root meaning to kindle, burn, or shine. It signifies the substance which was thought in ancient times to fill the upper regions of space, beyond the clouds.

Maxwell, Hertz, and Tesla

In 1864 James Clerk Maxwell of Edinburgh announced his electromagnetic theory of light, which was perhaps the greatest single step in the world's knowledge of electricity. Maxwell had studied and commented on the field of electricity and magnetism as early as 1855/6 when *On Faraday's lines of force* was read to the Cambridge Philosophical Society. The paper presented a simplified model of Faraday's work, and how the two phenomena were related. He reduced all of the current knowledge into a linked set of differential equations with 20 equations in 20 variables. This work was later published as *On Physical Lines of Force* in March 1861.

Around 1862, while lecturing at King's College, Maxwell calculated that the speed of propagation of an electromagnetic field is approximately that of the speed of light.

He considered this to be more than just a coincidence, and commented *"We can scarcely avoid the conclusion that light consists in the transverse undulations of the same medium which is the cause of electric and magnetic phenomena."* Working on the problem further, Maxwell showed that the equations predict the existence of waves of oscillating electric and magnetic fields that travel through empty space at a speed that could be predicted from simple electrical experiments; using the data available at the time, Maxwell obtained a velocity of 310,740,000 m/s.

In his 1864 paper *A Dynamical Theory of the Electromagnetic Field*, Maxwell wrote, *The agreement of the results seems to show that light and magnetism are affections of the same substance, and that light is an electromagnetic disturbance propagated through the field according to electromagnetic laws.*

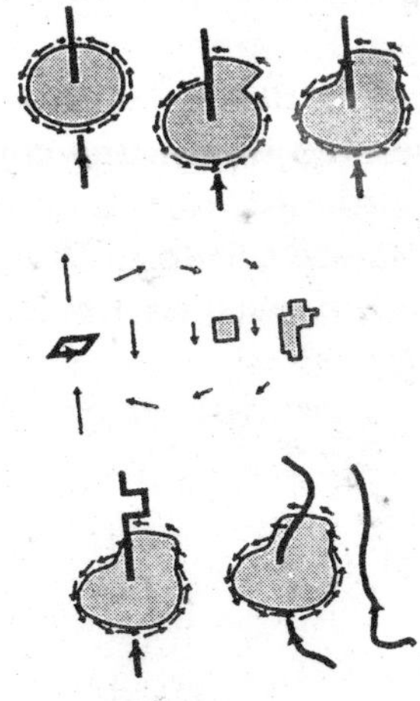

Fig. A Proof of Ampère's Law.

As already noted herein Faraday, and before him, Ampère and others, had inklings that the luminiferous ether of space was also the medium for electric action. It was known by calculation and experiment that the velocity of electricity was approximately 186,000 miles per second; that is, equal to the velocity of light, which in itself suggests the idea of a relationship between -electricity and "light."

A number of the earlier philosophers or mathematicians, as Maxwell terms them, of the 19th century, held the view that electromagnetic phenomena were explainable by action at a distance.

Maxwell, following Faraday, contended that the seat of the phenomena was in the medium. The methods of the mathematicians in arriving at their results were synthetical while Faraday's methods were analytical.

Faraday in his mind's eye saw lines of force traversing all space where the mathematicians saw centres of force attracting at a distance. Faraday sought the seat of the phenomena in real actions going on in the medium; they were satisfied that they had found it in a power of action at a distance on the electric fluids.

Both of these methods, as Maxwell points out, had succeeded in explaining the propagation of light as an electromagnetic phenomenon while at the same time the fundamental conceptions of what the quantities concerned are, radically differed. The mathematicians assumed that insulators were barriers to electric currents; that, for instance, in a Leyden jar or electric condenser the electricity was accumulated at one plate and that by some occult action at distance electricity of an opposite kind was attracted to the other plate.

Maxwell, looking further than Faraday, reasoned that if light is an electromagnetic phenomenon and is transmissible through dielectrics such as glass, the phenomenon must be in the nature of electromagnetic currents in the dielectrics.

He therefore contended that in the charging of a condenser, for instance, the action did not stop at the insulator, but that the "displacement" currents are set up in the insulating medium, which currents continue until the resisting force of the medium equals that of the charging force.

In a closed circuit conductor an electric current is also a displacement of electricity. The conductor offers a certain resistance, akin to friction, to the displacement, and heat is developed in the conductor, proportional as already stated herein to the square of the current, which current flows as long as the impelling electric force continues.

This resistance may be likened to that met with by a ship as in its progress it displaces the water. The resistance of the dielectric is of a different nature and has been compared to the compression of multitudes of springs, which, under compression, yield with an increasing back pressure, up to a point where the total back pressure equals the initial pressure. When the initial pressure is withdrawn the energy expended in compressing the "springs" is returned to the circuit, concurrently with the return of the springs to their original condition, this producing a reaction in the opposite direction.

Consequently the current due to the displacement of electricity in a conductor may be continuous, while the displacement currents in a dielectric are momentary and, in a circuit or medium which contains but little resistance compared with capacity or inductance reaction, the currents of discharge are of an oscillatory or alternating nature. Maxwell extended this view of displacement currents in dielectrics to the ether of free space.

Assuming light to be the manifestation of alterations of electric currents in the ether, and vibrating at the rate of light vibrations, these vibrations by induction set up corresponding vibrations in adjoining portions of the ether, and in this way the undulations corresponding to those of light are propagated as an electromagnetic effect in the ether. Maxwell's electromagnetic theory of light obviously involved the existence of electric waves in free space, and his followers set themselves the task of experimentally demonstrating the truth of the theory.

In 1887, Prof. Heinrich Hertz in a series of experiments proved the actual existence of such waves. The discovery of electric waves in space naturally led to the discovery and introduction in the closing years of the 19th century of wireless telegraphy, various systems of which are now in successful use on shipboard, lighthouses and shore and inland stations throughout the world, by means of which intelligence is transmitted across the widest oceans and large parts of continents. In 1891, notable additions to our knowledge of the phenomena of electromagnetic frequency and high potential current were contributed by Nikola Tesla.

Amongst the novel experiments performed by Tesla was to take in his hand a glass tube from which the air had been exhausted, then bringing his body into contact with a wire carrying currents of high potential, the tube was suffused with a pleasing bright glow.

Another experiment was to grasp a bulb that was suspended from a single wire attached to a high potential, high frequency current circuit, when a platinum button within the bulb was brought to vivid incandescence, the experimenter at this time standing on an insulating platform. The frequency and potential involved in the experiments made by Tesla at this time were of the order of one or more million cycles and volts. For further information relative to these experiments the reader may be referred to Tesla's Experiments with Alternate Currents of High Potential and High Frequency.

END OF THE CENTURY

The theories regarding electricity were undergoing change at the end of the 19 Century. Indeed it may with truth be said that the trend of all scientific investigation now leads to the conclusion that matter in its final analysis is electrical in its nature — in fact is electricity; the theory upon which this view is based being termed the electronic theory, or the electric theory of matter.

This theory(or better, hypothesis) in a word assumes that the atom of matter, so far from being indivisible, as assumed under the older theories, is made up of smaller bodies termed electrons, that these electrons arc electrical in their nature, and consequently all matter ultimately is electrical, the atoms of the different elements of matter consisting of a certain number of electrons, thus, 700 in the hydrogen atom and 11,200 in the oxygen atom.

This theory of matter in several of its important features is not altogether one of a day, nor is it due to the researches of one man or to the conception of one mind. Thus, as regards the view that the atom is not an indivisible particle of matter, but is made up of numerous electrons, many scientists have for years held that all the elements are modifications of a single hypothetical substance, protyle, "the undifferentiated material of the universe." Nor is the theory entirely new in its assumption that all matter is electrical. The electron as a unit of charge in electrochemistry was posited by G. Johnstone Stoney in 1874, who also coined the term *electron* in 1894.

Plasma was first identified in a Crookes tube, and so described by Sir William Crookes in 1879(he called it "radiant matter"). The place of electricity in leading up to the discovery of those beautiful phenomena of the Crookes Tube(due to Sir William Crookes), viz., Cathode rays, and later to the discovery of Roentgen or X-rays, must not be overlooked, since without electricity as the excitant of the tube the discovery of the rays might have been postponed indefinitely.

It has been noted herein that Dr. William Gilbert was termed the founder of electrical science. This must, however, be regarded as a comparative statement. During the late 1890s a number of physicists proposed that electricity, as observed in studies of electrical conduction in conductors, electrolytes, and cathode ray tubes, consisted of discrete units, which were given a variety of names, but the reality of these units had not been confirmed in a compelling way. However, there were also indications that the cathode rays had wavelike properties.

Faraday, Weber, Helmholtz, Clifford and others had glimpses of this view; and the experimental, work of Zecman, Goldstein, Crookes, J. J. Thomson and others had greatly strengthened this view. Over 35 years ago Weber predicted that electrical phenomena were due to the existence of electrical atoms, the influence of which on one another depended on their position and relative accelerations and velocities.

Helmholtz and others also contended that the existence of electrical atoms followed from Faraday's laws of electrolysis, and Johnstone Stoney, to whom is due the term "electron," showed that each chemical ion of the decomposed electrolyte carries a definite and constant quantity of electricity, and inasmuch as these charged ions are separated on the electrodes as neutral substances there must be an instant, however brief, when the charges must be capable of existing separately as electrical atoms; while in 1887, Clifford wrote: "There is great reason to believe that every material atom carries upon it a small electric current, if it does not wholly consist of this current."

In 1896 J.J. Thomson performed experiments indicating that cathode rays really were particles, found an accurate value for their charge-to-mass ratio e/m, and found that e/m was independent of cathode material. He made good estimates of both the charge e and the mass m, finding that cathode ray particles, which he called "corpuscles", had perhaps one thousandth of the mass of the least massive ion known(hydrogen). He further showed that the negatively charged particles produced by radioactive materials, by heated materials, and by illuminated materials, were universal. The nature of the Crookes tube "cathode ray" matter was identified by Thomson in 1897.

In the late 1800s, the Michelson-Morley experiment was performed by Albert Michelson and Edward Morley at what is now Case Western Reserve University.

It is generally considered to be the evidence against the theory of a luminiferous aether. The experiment has also been referred to as "the kicking-off point for the theoretical aspects of the Second Scientific Revolution." Primarily for this work, Albert Michelson was awarded the Nobel Prize in 1907.

Dayton Miller continued with experiments, conducting thousands of measurements and eventually developing the most accurate interferometer in the world at that time. Miller and others, such as Morley, continue observations and experiments dealing with the concepts. A range of proposed aether-dragging theories could explain the null result but these were more complex, and tended to use arbitrary-looking coefficients and physical assumptions.

By the end of the 19th century electrical engineers had become a distinct profession, separate from physicists and inventors. They created companies that investigated, developed and perfected the techniques of electricity transmission, and gained support from governments all over the world for starting the first worldwide electrical telecommunication network, the telegraph network.

Pioneers in this field included Werner von Siemens, founder of Siemens AG in 1847, and John Pender, founder of Cable & Wireless. The late 19th century produced such giants of electrical engineering as Nikola Tesla, inventor of the polyphase induction motor.

The first public demonstration of a "alternator system" took place in 1886. Large two-phase alternating current generators were built by a British electrician, J.E.H. Gordon, in 1882. Lord Kelvin and Sebastian Ferranti also developed early alternators, producing frequencies between 100 and 300 hertz. In 1891, Nikola Tesla patented a practical "high-frequency" alternator(which operated around 15,000 hertz).

After 1891, polyphase alternators were introduced to supply currents of multiple differing phases. Later alternators were designed for varying alternating-current frequencies between sixteen and about one hundred hertz, for use with arc lighting, incandescent lighting and electric motors. The possibility of obtaining the electric current in large quantities, and economically, by means of dynamo electric machines gave impetus to the development of incandescent and arc lighting.

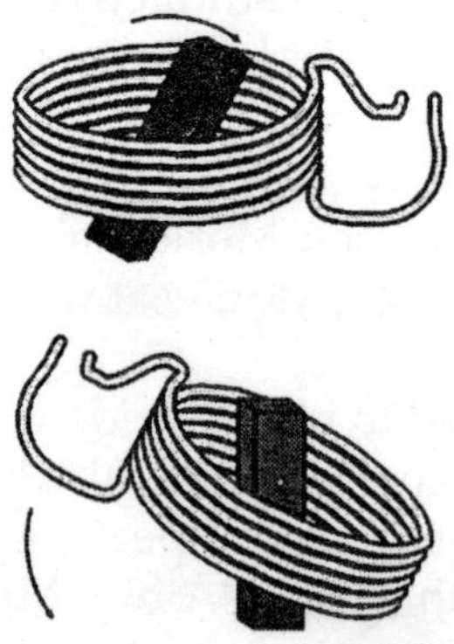

Fig. It doesn't Matter Whether it's the Coil or the Permanent Magnet that Spins. Either way, we get a Functioning Generator.

Until these machines had attained a commercial basis voltaic batteries were the only available source of current for electric lighting and power. The cost of these batteries, however, and the difficulties of maintaining them in reliable operation were prohibitory of their use for practical lighting purposes. The date of the employment of arc and incandescent lamps may be set at about 1877.

Even in 1880, however, but little headway had been made toward the general use of these illuminants; the rapid subsequent growth of this industry is a matter of general knowledge. The employment of storage batteries, which were originally termed secondary batteries or accumulators, began about 1879.

Such batteries are now utilized on a large scale as auxiliaries to the dynamo machine in electric power-houses and substations, in electric automobiles and in immense numbers in automobile ignition and starting systems, also in fire alarm telegraphy and other signal systems. The World's Columbian International Exposition was held in a building which was devoted to electrical exhibits.

General Electric Company had proposed to power the electric exhibits with direct current at the cost of one million dollars. However, Westinghouse, armed with Tesla's alternating current system, proposed to illuminate the Columbian Exposition in Chicago for half that price, and Westinghouse won the bid. It was an historical moment and the beginning of a revolution, as Nikola Tesla and George Westinghouse introduced the public to electrical power by illuminating the Exposition.

SECOND INDUSTRIAL REVOLUTION

The AC motor helped usher in the Second Industrial Revolution. The rapid advance of electrical technology in the latter 19th and early 20th centuries led to commercial rivalries. In the War of Currents in the late 1880s, George Westinghouse and Thomas Edison became adversaries due to Edison's promotion of direct current(DC) for electric power distribution over alternating current(AC) advocated by Westinghouse and Nikola Tesla. Tesla's patents and theoretical work formed the basis of modern alternating current electric power(AC) systems, including the polyphase power distribution systems.

Several inventors helped develop commercial systems. Samuel Morse, inventor of a long-range telegraph; Thomas Edison, inventor of the first commercial electrical energy distribution network; George Westinghouse, inventor of the electric locomotive; Alexander Graham Bell, the inventor of the telephone and founder of a successful telephone business.

In 1871 the electric telegraph had grown to large proportions and was in use in every civilized country in the world, its lines forming a network in all directions over the surface of the land. The system most generally in use was the electromagnetic telegraph due to S. F. B. Morse of New York, or modifications of his system. Submarine cables connecting the Eastern and Western hemispheres were also in successful operation at that time.

When, however, in 1918 one views the vast applications of electricity to electric light, electric railways, electric power and other purposes(all it may be repeated made possible and practicable by the perfection of the dynamo machine), it is difficult to believe that no longer ago than 1871 the author of a book published in that year, in referring to the state of the art of applied electricity at that time, could have truthfully written: "The most important and remarkable of the uses which have been made of electricity consists in its application to telegraph purposes". The statement was, however, quite accurate and perhaps the time could have been carried forward to the year 1876 without material modification of the remarks. In that year the telephone, due to Alexander Graham Bell, was invented, but it was not until several years thereafter that its commercial employment began in earnest. Since that time also the sister branches of electricity just mentioned have advanced and are advancing with such gigantic strides in every direction that it is difficult to place a limit upon their progress. For a more adequate account of the use of electricity in the arts and industries.

AC replaced DC for central station power generation and power distribution, enormously extending the range and improving the safety and efficiency of power distribution. Edison's low-voltage distribution system using DC ultimately lost to AC devices proposed by others: primarily Tesla's polyphase systems, and also other contributors, such as Charles Proteus Steinmetz(in 1888, he was working in Pittsburgh for Westinghouse).

The successful Niagara Falls system was a turning point in the acceptance of alternating current. Eventually, the General Electric Company(formed by a merger between Edison's companies and the AC-based rival Thomson-Houston) began manufacture of AC machines. Centralized power generation became possible when it was recognized that alternating current electric power lines can transport electricity at low costs across great distances by taking advantage of the ability to change voltage across the distribution path using power transformers. The voltage is raised at the point of generation(a representative number is a generator voltage in the low kilovolt range) to a much higher voltage(tens of thousands to several hundred thousand volts) for primary transmission, followed to several downward transformations, to as low as that used in residential domestic use.

The International Electro-Technical Exhibition of 1891 featuring the long distance transmission of high-power, three-phase electrical current. It was held between 16 May and 19 October on the disused site of the three former "Westbahnhöfe"(Western Railway Stations) in Frankfurt am Main. The exhibition featured the first long distance transmission of high-power, three-phase electrical current, which was generated 175 km away at Lauffen am Neckar. As a result of this successful field trial, three-phase current became established for electrical transmission networks throughout the world.

Much was done in the direction in the improvement of railroad terminal facilities, and it is difficult to find one steam railroad engineer who would have denied that all the important steam railroads of this country were not to be operated electrically. In other directions the progress of events as to the utilization of electric power was be expected to be equally rapid.

In every part of the world the power of falling water, nature's perpetual motion machine, which has been going to waste since the world began, is now being converted into electricity and transmitted by wire hundreds of miles to points where it is usefully and economically employed. The extensive utilization of falling water was not limited to natural water falls. In hundreds of places where a fall of 40 to 400 feet extends over 10 to 50 miles, and where in the aggregate hundreds of thousands of horse power, by suitable hydraulic methods, are available, the power was usefully employed, thereby in large measure conserving the limited quantity of the world's coal.

It has for instance been proposed to dam Niagara River at the foot of the gorge whereby another source of water power equal to that at the present falls would be available.

The Jchlun River in Kashmir, India, too, has a fall of 2,480 feet in 80 miles with a minimum flow of 30,000 gallons per second, and a beginning has been made to develop the 1,000,000 electric horse power here represented, ÿ considerable portion of which it is proposed to utilize in the production of nitrate of lime for fertilizer purposes, by combining by means of powerful electric currents the limestone that abounds in this region with the nitrogen of the air, a combination which Danish engineers have shown to be commercially possible, and which inexhaustible product may in time be economically available to replenish the failing powers of the farm lands of America and other countries.

The dreams of the electrical engineer were that the direct

production of electricity from coal without the intervention of the steam engine with its wasteful methods was to be realized.

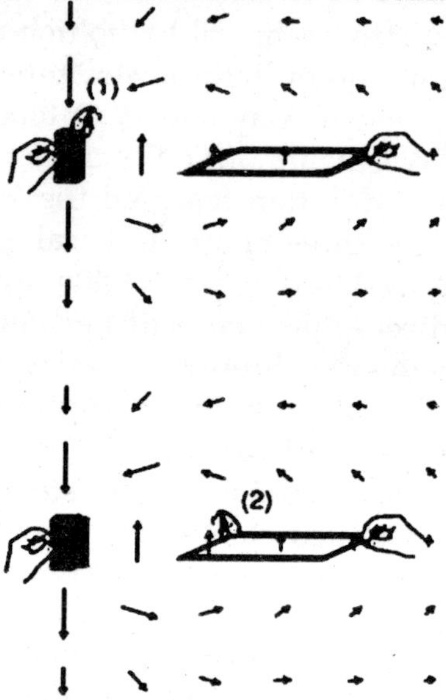

Fig. A Generator that Works with Linear Motion.

The first windmill for electricity production was built in Scotland in July 1887 by Prof James Blyth of Anderson's College, Glasgow(the precursor of Strathclyde University. Across the Atlantic, in Cleveland, Ohio a larger and heavily engineered machine was designed and constructed in 1887-1888 by Charles F. Brush, this was built by his engineering company at his home and operated from 1886 until 1900. The Brush wind turbine had a rotor 56 feet(17 m) in diameter and was mounted on a 60 foot(18 m) tower.

Although large by today's standards, the machine was only rated at 12 kW; it turned relatively slowly since it had 144 blades. The connected dynamo was used either to charge a bank of batteries or to operate up to 100 incandescent light bulbs, three arc lamps, and various motors in Brush's laboratory. The machine fell into disuse after 1900 when electricity became available from Cleveland's central stations, and was abandoned in 1908.

20TH CENTURY

Various units of electricity and magnetism have been adopted and named by representatives of the electrical engineering institutes of the world, which units and names have been confirmed and legalized by the governments of the United States and other countries.

Thus the Volt, from the Italian Volta, has been adopted as the practical unit of electromotive force, the Ohm, from the enunciator of Ohm's law, as the practical unit of resistance; the Ampere, after the eminent French scientist of that name, as the practical unit of current strength, the Henry as the practical unit of inductance, after Joseph Henry and in recognition of his early and important experimental work in mutual induction.

Lorentz and Poincaré

Between 1900 and 1910, many scientists like Wilhelm Wien, Max Abraham, Hermann Minkowski, or Gustav Mie believed that all forces of nature are of electromagnetic origin(the so called "electromagnetic world view"). This was connected with the electron theory developed between 1892 and 1904 by Hendrik Lorentz.

Lorentz introduced a strict separation between matter(electrons) and ether, whereby in his model the ether is completely motionless, and it won't be set in motion in the neighborhood of ponderable matter. Contrary to other electron models before, the electromagnetic field of the ether appears as a mediator between the electrons, and changes in this field can propagate not faster than the speed of light. Lorentz theoretically explained the Zeeman effect on the basis of his theory, for which he received the Nobel Prize in Physics in 1902.

A fundamental concept of Lorentz's theory in 1895 was the "theorem of corresponding states" for terms of order v/c. This theorem states that a moving observer(relative to the ether) in his "fictitious" field makes the same observations as a resting observer in his "real" field. This theorem was extended for terms of all orders by Lorentz in 1904. Lorentz noticed, that it was necessary to change the space-time variables when changing frames and introduced concepts like physical length contraction(1892) to explain the Michelson-Morley experiment, and the mathematical concept of local time(1895) to explain the aberration of light and the Fizeau experiment.

That resulted in the formulation of the so called Lorentz transformation by Joseph Larmor(1897, 1900) and Lorentz(1899, 1904).

Continuing the work of Lorentz, Henri Poincaré between 1895 and 1905 formulated on many occasions the Principle of Relativity and tried to harmonize it with electrodynamics. He declared simultaneity only a convenient convention which depends on the speed of light, whereby the constancy of the speed of light would be a useful postulate for making the laws of nature as simple as possible. In 1900 he interpreted Lorentz's local time as the result of clock synchronization by light signals, and introduced the electromagnetic momentum by ascribing to electromagnetic energy the "fictitious" mass $m = E/c$.

And finally in June and July 1905 he declared the relativity principle a general law of nature, including gravitation. He corrected some mistakes of Lorentz and proved the Lorentz covariance of the electromagnetic equations.

Poincaré also found out that there exist non-electrical forces to stabilize the electron configuration and asserted that gravitation is a non-electrical force as well. So the electromagnetic world view was shown by Poincaré to be invalid. However, he remained the notion of an ether and still distinguished between "apparent" and "real" time and therefore failed to invent what is now called special relativity.

Einstein's Annus Mirabilis

In 1905, while he was working in the patent office, Albert Einstein had four papers published in the *Annalen der Physik,* the leading German physics journal. These are the papers that history has come to call the *Annus Mirabilis Papers*:

- His paper on the particulate nature of light put forward the idea that certain experimental results, notably the photoelectric effect, could be simply understood from the postulate that light interacts with matter as discrete "packets"(quanta) of energy, an idea that had been introduced by Max Planck in 1900 as a purely mathematical manipulation, and which seemed to contradict contemporary wave theories of light. This was the only work of Einstein's that he himself called "revolutionary."
- His paper on Brownian motion explained the random movement of very small objects as direct evidence of molecular action, thus supporting the atomic theory.
- His paper on the electrodynamics of moving bodies introduced the radical theory of special relativity, which showed that the observed independence of the speed of light on the observer's state of motion required fundamental changes to the notion of simultaneity. Consequences of this include the time-space frame of a moving body slowing down and contracting(in the direction of motion) relative to the frame of the observer. This paper also argued that the idea of a luminiferous aether—one of the leading theoretical entities in physics at the time—was superfluous.
- In his paper on mass–energy equivalence(previously considered to be distinct concepts), Einstein deduced from his equations of special relativity what later became the well-known expression: $E = mc$, suggesting that tiny amounts of mass could be converted into huge amounts of energy.

All four papers are today recognized as tremendous achievements—and hence 1905 is known as Einstein's "Wonderful Year". At the time, however, they were not noticed by most physicists as being important, and many of those who did notice them rejected them outright. Some of this work—such as the theory of light quanta—remained controversial for years.

Chapter 3

Electromagnetic Waves

Electromagnetic interactions between charged particles propagate at a large but finite speed, the speed of light. Jiggle a charge over there, a charge here won't react, won't feel a force change, until the pulse reaches it. This is what gives such prominence and reality to the electric and magnetic field concepts, even though from the point of view of forces between material particles the fields may seem to be mere middlemen: charge produces field, field exerts force on another charge.

Maxwell's equations are expressed in terms of these middlemen. There are infinitely many different solutions of Maxwell's equations. For example, for a pulse traveling in free space along the +*x* direction the solutions for the E and B fields have the form

$$\mathrm{E} = \mathrm{E}_0 F(x - ct);\ \mathrm{B} = \mathrm{B}_0 F(x - ct),$$

where c is the speed of light, E_0 is a constant vector of arbitrary magnitude perpendicular to the x axis and B_0 is a constant vector perpendicular to both E_0 and the x axis. In cgs units these two vectors must have the same magnitude.

In Equation F is an arbitrary function of the argument indicated. It should be evident, just from the fact that F depends on x and t only in the combination $x - ct$, that the pulse travels at speed c to the right along the x axis, preserving its shape.

There are other solutions that describe a pulse moving to the left, in the direction of the negative x axis. They have the same structure as above but with $F(x - ct)$ replaced by $G(x + ct)$, where G is again an arbitrary function, now of the combination $x + ct$. And there are analogous solutions for pulses traveling in all other directions. The nature of Maxwell's equation is such that, given any particular set of solutions, their sum is also a solution!

Let us return to the case of propagation along the +*x* axis and the function $F(x - ct)$ encountered there. A special case is the sinusoidal function, $F(x - ct) = \sin\{\mathrm{k}(x - ct) + \phi\}$, [1.2] where φ is an arbitrary "phase" constant and k an arbitrary "wave number" constant. Recall that the sine function and its derivative repeat when the argument is augmented by any positive or negative multiple of 2π.

Thus, for given time t the signal repeats when x goes from x_1 to x_2 provided that $k(x_2 - x_1) = 2\pi$(we measure angles in *radians*; 2π radians $= 360°$). The repetition distance $x_2 - x_1 = \lambda$ defines the *wavelength*; hence we identify k with the reciprocal wavelength according to $k = 2\pi/\lambda$. Similarly, for given position x the signal repeats itself in a time interval such that $kc = 2\pi$.

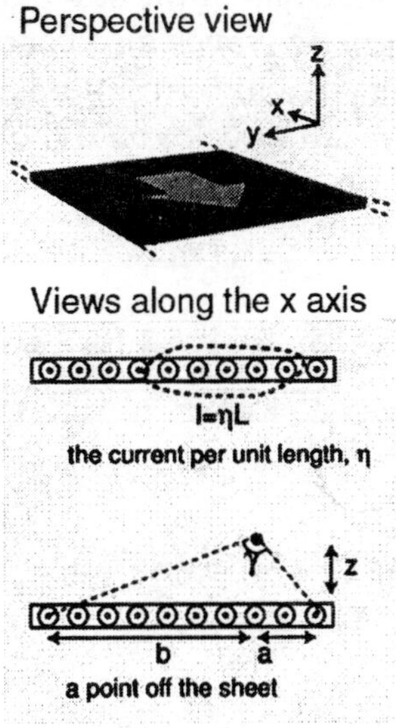

Fig. A Sheet of Charge.

This time interval is the *period* of the oscillatory signal. Its reciprocal is the repetition frequency f. Thus, $f = kc/2\pi$. We hereby recapture the familiar high-school formula $f\lambda = c$: the product of frequency and wavelength is equal to the speed of light.

In(almost) all that follows, in order to avoid too many writings of 2π we will use the so-called *circular* frequency ω, defined by $\omega = 2\pi f$. The circular frequency is 2π times the conventional repetition frequency f. The wave number k is just 2π divided by the wavelength. The circular frequency and wave number are related by $\omega = kc$.

The most general function $F(x - ct)$ describing a signal propagating to the right along the x axis is a superposition of the sinusoidal solutions given above, summed over all wave numbers, with the phase φ and amplitudes E_0 and B_0 chosen independently for each wave number(but with $|E_0| = |B_0|$).

A completely general solution of the free space Maxwell equations is a superposition of this kind of superposition, taken over all *directions* of propagation! The radiation coming out of the sun or a light bulb involves just such a superposition, with a range of wavelengths mostly concentrated in the visible wavelength region 0.4—0.7 microns(1 micron = 10″ cm). Our eyesight, of course, evolved to respond over this interval and light bulbs are designed to accommodate to our eyes, more or less.

We must also remark here that electromagnetic waves carry energy; they cause material charges to jiggle and hence acquire kinetic energy. We would not be here if the sun's rays did not carry energy to the earth. Electromagnetic waves also carry momentum, though this is less familiar in everyday life. An intense enough beam of light can not only warm you, it can knock you over.

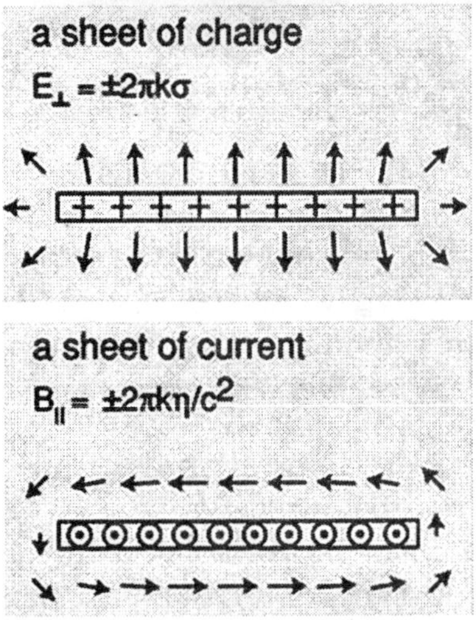

Fig. A Sheet of Charge and a Sheet of Current.

BLACKBODY RADIATION

It has been known since antiquity that when metals and other substances are heated to high enough temperatures, they radiate visible light; the higher the temperature, the bluer the light. The reasons became at least qualitatively clear in the mid nineteenth century in connection with the developing understanding both of thermodynamics and of electromagnetism.

Light is nothing but an electromagnetic disturbance that is generated by the jiggling of charges and that propagates through space. Higher temperature implies increased jiggling; hence greater radiation intensity and also, it happens, a shift toward higher frequencies. In the 1850s Gustav Kirchhoff, a master of both of the above-mentioned disciplines, reasoned his way to a critical finding. Consider a hollow vessel whose walls are maintained at some temperature T. It was expected that the walls must be capable of both emitting and absorbing electromagnetic radiation.

Although the atomic picture was not well developed at that time, one knew that electric charge is somehow present in matter and that the jiggling of electric charge must lead to *emission* of radiation.Conversely, incident radiation induces jiggling, which leads to *absorption* of energy from the radiation. Reflecting a balance between emission and absorption, the hollow vessel will be filled with electromagnetic radiation, with waves moving in every possible direction and covering a whole spectrum of frequencies.

By simple but ingenious thermodynamic reasoning Kirchhoff could show that the radiation intensity must be isotropic(rays moving equally in all directions) and uniform over the vessel(the same intensity here as there).

More strikingly, he could also show that the spectrum of the radiation, the radiation energy density as a function of frequency, must be completely independent of the material of which the walls are made. Let u be the radiation energy density(energy per unit volume) in a unit frequency interval at frequency ω.

Since u does not depend on the nature of the walls, on all those detailed material parameters that it *might* have depended on, it must be a universal function $u = u(\varphi, T)$ of frequency and temperature only. Precisely because it is universal this "blackbody" spectral function must be something of fundamental interest, something to be not only tackled experimentally but understood theoretically. It took about forty years for that understanding to emerge; or rather, to begin to emerge.

As said earlier, it was the German physicist Max Planck who did it. That was in the year 1900. But first let's consider a few things that came before. Some years earlier, the Austrian experimentalist Josef Stefan had discovered experimentally that the total energy density—the energy density u integrated over all frequencies—is proportional to the fourth power of the absolute temperature T. Later, Ludwig Boltzmann could prove this on purely thermodynamic grounds.

In 1893 W. Wien proved, again by a beautiful thermodynamic argument, that u(w, T) must have the form $u = \omega^3 W(\omega / T)$ where W is some function of the ratio indicated, a function that Wien could however not predict theoretically. The reasoning that led to the above formula was impeccable. A few years later Wien reasoned himself to another result, this one not quite so impeccable; namely, $W(\omega / T) = A\exp(-b\omega / T)$, where A and b are unspecified constants. In the middle of 1900, Lord Rayleigh(William Strutt) revisited the whole problem, making better use of the doctrines of statistical mechanics that had been developing all the while. He came up with the disastrous result $u = k_B T\omega^2 / \pi^2 c^3$, where k_B here is a statistical mechanics parameter named after Ludwig Boltzmann. Rayleigh's result was disastrous because the predicted energy density integrated over all frequencies, the total energy density, is infinite! Rayleigh made excuses and dropped the subject.

On October 7, 1900 in Berlin, the Plancks entertained Mr. and Mrs. H. Rubens for tea. Rubens was Planck's colleague, an experimentalist who had been carrying out measurements of the blackbody spectrum. Amidst the socializing Rubens drew Planck aside and showed him his latest results.

That very evening Planck, who had been brooding over the blackbody problem for some time, sat down and worked out an empirical formula that interpolated between low frequencies, where Rayleigh's formula seemed to fit, and very high frequencies, where Wien's seemed to fit. Planck's formula agreed beautifully with the data in between as well! Both Rubens' experimental results and Planck's formula were announced in Berlin within a fortnight.

Planck was a master of thermodynamics, though a conservative fellow who was rather skeptical about the newfangled statistical mechanics. To his everlasting credit he did not rest on his empirical success. He set out to derive his formula from first principles.

Luckily, he seems not to have been aware of Rayleigh's disastrous result, which was unavoidable within the classical framework of the times. Planck took a more complicated path. Since the radiant energy function is independent of the nature of the vessel walls, he was free to assume that the walls consist of simple oscillators, charged particles at the ends of springs, with all possible spring frequencies represented.

By impeccable electromagnetic arguments he could relate the spectral function $u(\omega, T)$ to the thermodynamic mean energy $E(\omega, T)$ of a spring of frequency ω. Had he obtained the right classical result for this latter energy, he would have ended up with Rayleigh's formula.

Instead, he fiddled around, then introduced a rather arbitrary assumption which he later acknowledged was done in desperation to achieve the result he wanted. He supposed that the oscillator can take on only those energy values ε that are integer multiples of a constant times frequency: $\varepsilon = n\hbar\omega$, where n is any non-negative integer.

In effect, on this model the walls could radiate and absorb only in packets of energy $\hbar\omega$. The proportionality constant $\hbar$ is what we shall here call Planck's constant. But since Planck used repetition frequency

f rather than circular frequency ω, he wrote = nhf, so *our* Planck's constant $\hbar$ is related to Planck's Planck constant by $\hbar = h/2\pi$. Of course Planck did not predict the numerical value of his constant. It entered the world as a new parameter of nature. Planck's blackbody formula in our current notation is

$$u = \frac{\hbar\omega^3}{\pi^2 c^3} \frac{1}{\exp(\hbar\omega / k_b T) - 1}.$$

Fitting this to the available data he could determine both the constant ' and Boltzmann's constant k_B. Knowing the latter he could through well-established arguments determine the number of molecules in a mole as well as the electric charge on the electron! The results were quite good. The modern value of Planck's constant is

$$\hbar = 1.055 \times 10^{-27} \text{ erg-sec} = 6.58 \times 10^{-16} \text{eV-sec}.$$

The erg is the unit of energy in the cgs system. One food calorie is worth about 40 billion ergs! As noted earlier, the symbol eV stands for electron volt, another common unit. Notice that Planck's constant has the dimensions of energy × time; equivalently, of momentum × length.

As we now know, the whole universe is filled with blackbody radiation left over from the Big Bang. It has cooled down at our epoch to the low temperature of 2.7 degrees above absolute zero.

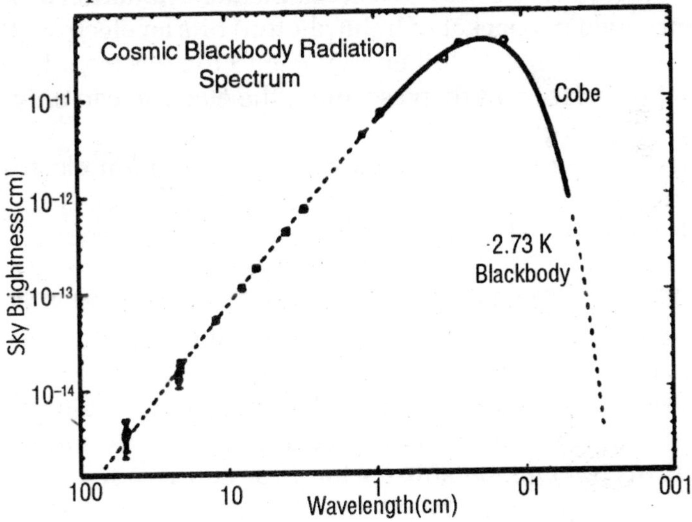

Fig. The Spectrum of Radiation in the Cosmos left over from the Big Bang, Plotted as a Function of Wavelength. The Solid Curve and Boxes come from Experiment(the Cosmic Background Explorer, COBE); the Dotted line is the Theoretical Blackbody Curve for Temperature $T = 2{:}73\ K$ above Absolute Zero. The fit is Quite Spectacular.

Figure given above displays the experimental findings and the theoretical blackbody curve(dotted line) corresponding to the current cosmic temperature. Closer in time to the Big Bang, the temperature was very much higher indeed.

Planck had clearly obtained a remarkably good fit to the data but it was not clear whether he had really broken new ground. Statistical mechanics was still on an unsure footing. It was Einstein above all who recognized that a revolution was at hand.

Planck and others thought that the new, funny business reflected something peculiar about the interaction of charged particles and radiation. Einstein in 1905 saw more deeply.

This phenomenon of energy packets, he said, is intrinsic to the radiation itself; intrinsically, radiation of frequency → can exist only in packets of energy $\hbar\,\omega$. He proposed a test. It had been known for some years that charged particles come off a metallic surface when it is irradiated with ultraviolet light. J. J. Thomson had verified that these particles are electrons. The current of outgoing electrons was known to increase with the intensity of the radiation.

No surprise in that. But one would have thought that the energy of the electrons would also increase with radiation intensity. Einstein said otherwise.

Whatever may be the intensity of incident radiation of some given frequency, when a packet of light(photon) hits an electron, the photon can at most transfer its full energy $\hbar\,\omega$ to the electron. In making its way to the surface and then escaping, the electron can only lose some part of that energy.

Einstein therefore predicted that the maximum electron energy, independent of the intensity of the incident radiation, is $E_{max} = \hbar\,\omega - \Phi$. Here Φ is the metal's work function, the energy required to escape from the surface. This "photoelectric" formula was not well tested until some years later, starting with the experiments of O. W. Richardson in 1912; then those of K. T. Compton, R. A. Millikan, and others.

Einstein's notion of energy packets was initially received with considerable skepticism, although he became increasingly respected after 1905 for his work on relativity and Brownian motion. When he was put up for membership in the Prussian Academy, his backers, including Planck, said *à propos* the packets that some excuses had to made for such an otherwise productive fellow.

Einstein knew from the beginning that in a directed beam of light the packets carry not only energy but also momentum p of magnitude $p = \hbar\,\omega/c = 2\pi\hbar\,'/\lambda$. These packets are like particles, carrying energy and momentum.

As particles they seemed quite unusual: they are massless and therefore always travel at the speed of light, whatever the energy. They later came to be called *photons*. The real clincher came with A. H. Compton's 1922 paper reporting experiments on the scattering of X-rays off electrons. The scattering reaction is $\gamma + e \to \gamma + e$, where γ denotes a photon. The experimentally observed kinematics was in every respect that describing a zero-mass particle scattering off an electron.

The big trouble with all of this, however, was that light was known to be a wavelike phenomenon. How could it also display these particle-like properties? This was the great wave particle duality conundrum. It plagued everybody who thought about it, especially Einstein.

EARLY SPECTROSCOPY

It had been known since antiquity that familiar sources of light—the sun, flames, heated substances of any kind—emit a mixture of colors of light or, as we would now say, a mixture of frequencies. The rainbow is a familiar case in which the spectrum is spread out by natural means. With Newton's prisms one learned that these colour mixtures can be separated at will, whatever the source of light. One speaks of the *spectrum* of radiation emitted by a source, the intensity as a function of frequency. We are not now restricting ourselves to blackbody radiation but are considering radiation sources more generally. The spectrum of any source will depend on the nature of the emitting material and on the material's condition, thermal and otherwise.

In general, cold stuff doesn't radiate much at all. Radiation intensity increases with temperature. But a sample of matter can be stimulated into radiative ferment by other means as well; for example, by zapping it with an electrical spark, bombarding it with a beam of fast particles, and so on.

The spectrum will inevitably be spread out continuously over frequency, but there will also often be sharp peaks in intensity centered around certain particular frequencies. Because of the way spectral data is usually presented in charts, these peaks are called "lines." The discovery and beginning studies of spectral lines date back to the early 1800s. Actually, depending on circumstances, there can be dark lines superimposed on a continuum as well as bright ones. The bright ones represent enhanced emission at certain special frequencies. The dark ones represent enhanced absorption of radiation that is seeking to emerge from lower layers of the material.

In either case, the line spectrum differs from one species of atom or molecule to another. Indeed, a new set of lines not previously known on earth was first discovered in the solar spectrum and only later identified with helium, subsequently discovered here on earth. Early interest in spectroscopy was largely centered on its role in chemical identification and discovery.

But it also seemed to some that the lines could serve as messengers from inside the atom, that they ought to have a role in telling us about those insides. The prevailing view in the nineteenth century was that the lines correspond to the frequencies of various modes of oscillation of electric charge within the atom. According to classical electromagnetism, oscillating charges can produce and absorb radiation.

Relatedly, the thought was that each atom radiates at all its characteristic frequencies at once. Spectroscopists began to look at the data in a purely empirical spirit to see if they could spot any regularities in the line frequencies; for example, evidence that the line frequencies are simple harmonics of a fundamental frequency characteristic of the particular species of atom.

That latter idea was not borne out. What did prove to be a fateful discovery was made by Johann Balmer(1825—98), aged about 60 at the time, a teacher at a Swiss girls' school, a man who had never before published a single paper in physics and whose main interest seems to have been in architecture.

As did others before him, he thought that the spectrum of the hydrogen atom might be the best place to look for regularities. He took up data from the work of A. Angström, who had discovered four lines in the visible part of the hydrogen spectrum and who had measured their wavelengths γ with impressive accuracy.

Balmer could fit the data with the remarkably simple formula

$$\gamma = \text{constant} \times \frac{m^2}{m^2 - 2^2}, \; m = 3, 4, 5, 6.$$

With that single adjustable constant in front, the formula worked well for all four lines. In a subsequent paper, having learned of more recent results on other lines, Balmer could get an excellent fit for lines corresponding to m ranging up to $m = 14$.

Others now took up the game for multielectron atoms, trying various formulas, with only moderate success. But in the early 1900s one idea emerged that did prove to be quite fruitful.

It was suggested that one should look for formulas in which the line *frequencies* are expressed as *differences* of simple expressions. This was the idea especially of W. Ritz and came to be known as the Ritz combination principle. Actually, instead of frequency, consider what is the same variable up to a multiplicative constant, the inverse wavelength. Then notice that for hydrogen, the Balmer formula becomes

$$\frac{1}{\gamma} = \text{constant} \times \left(\frac{1}{2^2} - \frac{1}{m^2}\right),$$

a difference indeed between two very simple expressions.

THE RUTHERFORD ATOM

Early in the first decade of the twentieth century, Ernest Rutherford was ensconced at Manchester, studying the passage of alpha particles through thin metal foils. Recall that the α particle is the nucleus of the helium atom.

It was known in Rutherford's time that energetic α particles are ejected in the radioactive decay of certain atoms. That was interesting in its own right, but it also provided a source of energetic particles with which to bombard and thereby probe the structure of atoms. As was expected from then-current models of the atom, Rutherford found that the á particles typically scatter only through very small angles in passing through a thin metal foil.

He set his colleagues Geiger and Marsden to investigate the possibility that there might occasionally be rare, large-angle events, scatterings through even more than 90°. There were such events! Not a lot, but many more than would have been expected. Rutherford was astonished. He sat down, thought, calculated, and came up with a revolutionary new picture of the atom.

It was impossible; he reasoned correctly, that the large-angle events could have been caused by scattering off electrons. The electron mass is much too small to enable it to seriously deflect the much heavier α particle. The large-angle events must therefore arise from scattering off a more massive object in the atom, presumably the object that contains the positive charge of the atom.

From the kinematics of such a collision he could show that in order to account for the large-angle events, the target must have a mass bigger than that of the á particle. Also, it must be very small in size, so that when the á particle comes close it can feel a strong enough repulsive Coulomb force to be seriously turned off course. Indeed, the radius cannot be much larger than about 10" cm, he deduced.

All of this was for gold leaf film. The size of the whole atom was known from other considerations to be roughly 10– cm. The central, positive body—the nucleus—was therefore so small that for analysis of scattering it could be treated as a point. Rutherford worked out a formula for the expected distribution in scattering angles using purely classical dynamics. The answer depends on the charge to mass ratio of the á particle, which was known well enough, and on the charge *Ze* of the nucleus, which was not well known. The *shape* of the experimental curve fit the theory quite successfully. The absolute level was off. As we now know, Rutherford was off by almost a factor of 2 in the Z value for gold. But never mind. His model was a winner.

There was a remarkable piece of luck in all of this. Scattering, like all else, is governed by quantum mechanical rather than classical Newtonian laws. For most phenomena at the atomic level the two doctrines produce quite different predictions.

It just so happens that for scattering in a Coulomb force field the two agree in very good approximation. Rutherford's classical reasoning produced the right scattering formula and led to the right picture of the atom.

The Rutherford atom can be pictured as a kind of solar system, with all the positive charge concentrated in a nucleus that is tiny in size but that contains essentially all the mass of the atom. The electrons travel in orbits about the nucleus. The radius of the nucleus depends on the atomic species in question, but as we now know it is in fact of order 10–*12* cm.

BOHR'S QUANTUM MODEL

Despite its immediate appeal, the Rutherford atom faced some very big hurdles, as indeed did any of the atomic models that had preceded it. Let us illustrate these problems on the example of hydrogen, the simplest of neutral atoms.

The nucleus of the hydrogen atom is a single proton. Its charge is balanced by a single orbiting electron. The electron is in a state of acceleration as it moves about the nucleus, since it is continually being acted on by the Coulomb force of the nucleus. According to classical electromagnetism, an accelerating charge emits electromagnetic radiation. Suppose for a moment that we can ignore the fact that the electron must therefore be continually losing energy. We will come back to that. For the moment let it radiate, but ignore the loss of energy.

One can then easily work out the orbital dynamics. The orbits are ellipses, of which the circle is a special case. The motion around an ellipse is of course periodic in time. According to classical electrodynamics, a charge undergoing periodic motion will radiate at the frequency of that orbital motion.

The frequency depends on orbit parameters. But given any macroscopic collection of atoms, one would expect to find an essentially continuous range of orbit parameters. It is classically incomprehensible that the electrons would select only certain orbits and not others. It is therefore hard to see why only a discrete set of lines is observed. Anyhow, precisely because it is radiating we cannot ignore the fact that the electron is continually losing energy. This means that it must eventually spiral into the nucleus, whirling faster and faster on the way in and thereby producing a continuous spectrum. So, why don't atoms collapse? What stabilizes them? And again, why do they radiate at only certain frequencies?

Along comes the young Danish student Niels Bohr, on a stay in Cambridge to work with J. J. Thomson. Thomson had his own model of the atom, a scheme now remembered chiefly by historians. Bohr was critical of it, very politely so, but critical.

In 1912 he moved on to Manchester to work with Rutherford. And there his great idea came to him. Some of the views that Bohr developed had been suggested by others around that time, but it was Bohr who had the unerring instinct for the right path. The general view in the late nineteenth century was that the atom must have many modes of classical vibration, and that each atom radiates at all its characteristic frequencies simultaneously.

But by the early years of the new century, an alternative idea was suggested; namely, that at any given moment an atom radiates at only one or another of its characteristic frequencies, and that the whole spectrum of lines from a bulk sample of atoms arises because different atoms are radiating different lines at any given moment.

Bohr adopted this picture. He also firmly adopted the view that Planck's quantum must somehow enter into the atomic story. That may seem obvious in retrospect but it was not obvious at the time. After all, most of physics was going along merrily in a classical mode alongside the limited quantum forays initiated by Planck, Einstein, and a few others. But Bohr thought that the quantum must be essential for an understanding of the stability of the atom. What he did for the one-electron atom can be described in terms of the following steps.

- At the start, simply forbid the electron to radiate; and calculate the electron orbit on purely classical grounds. Since the nuclear Coulomb force obeys an inverse square law, the dynamical problem is the same as for the motion of planets around the sun. One knows all about the motion. The orbits are ellipses. Following Bohr, let us in particular specialize to circular orbits, where the arithmetic is especially easy. With Ze the nuclear charge and with the nucleus treated as a point particle(which it essentially is on the scale of the whole atom), the attractive radial force on the electron is

$$F(r) = -Ze/r.$$

The potential energy corresponding to this attractive force is

$$V(r) = -Ze/r.$$

The(inward) acceleration of a particle traveling at velocity v in a circular orbit is $a = v/r$. From Newton's law, therefore, $mv^2 = Ze^2/r$ The(nonrelativistic) energy, kinetic plus potential, is

$$E = mv^2/2 + V(r) = -Ze^2/2r$$

The angular velocity is $\omega = v/r$. Finally, let's introduce the angular momentum L, a vector quantity defined quite generally by

$$L = mr \times v.$$

For a circular orbit the position and velocity vectors are perpendicular to one another, so L points in a direction perpendicular to the plane of motion. Its magnitude is $L = mrv$ The five variables r, v, E, ω, and L are related through the above four equations. If we know any one of these quantities, we know the others. Let us single out L and express all the others in terms of it. One can easily check that

$$r = \frac{L^2}{Zme^2}, v = \frac{Ze^2}{L}; \omega = \frac{Z^2me^4}{L^3}; E = -\frac{Z^2me^4}{2L^2}.$$

Classically, of course, L can take on values ranging *continuously* from zero to infinity.

- In this step we will take some liberties with history, focusing on only one of several lines of argument that Bohr used to motivate a revolutionary fi quantum condition fl that he introduced. Out of the blue, more or less, Bohr postulated that L can take on only a discrete set of values,

$$L = n\hbar,$$

where n ranges over the positive integers, $n = 1, 2, 3, \ldots \infty$ The circular orbits, labeled by the integer n, are hereby quantized by fiat! For the nth orbit it now follows that the radius, velocity, angular velocity, and energy are all similarly quantized, with

$$r_n = \frac{n^2}{Z}\left(\frac{\hbar^2}{me^2}\right); v_n = \frac{Z}{n}\left(\frac{e^2}{\hbar c}\right)c;$$

$$\omega_n = \frac{Z^2}{n^3}\left(\frac{me^4}{\hbar^3}\right); E_n = -\frac{Z^2}{n^2}\left(\frac{me^4}{2\hbar^2}\right).$$

The natural length in this problem is the *Bohr radius,*

$$a_B = \hbar / me = 0.53 \text{ angstroms},$$

where 1 angstrom = 10– cm. The natural energy is the *Rydberg,*

$$RY = me/2\hbar = e/2\alpha_{B;}$$

numerically, 1Ry = 13.6 electron volts. Finally,

$$\alpha = e/\hbar c = 1/137$$

is the so-called *fine structure constant*. The integer n is often called the *principal quantum number.*

- Having ignored radiation and imposed his quantum condition to determine the allowed circular orbits, Bohr now asserted that radiation is emitted when, and only when, the electron "decides" to jump down from an orbit of energy E_n to one of lower energy $E_{n¢}$. When this occurs, radiation of frequency w_g is emitted, the photon carrying away the energy difference:

$$\hbar\omega_\gamma = E_n - E_{n'}.$$

Conspicuously, Bohr does not tell us how and when the electron decides to jump in the process of emission. In addition to emission of radiation, there is also the phenomenon of absorption. The atom can absorb an incident photon of the right frequency by jumping *up* from one level to another of higher energy. The incident photon energy has to be just such as to supply the energy difference between the two electron levels.

Bohr's allowed states of motion(allowed orbits) are often called "stationary states," to emphasize that(by Bohr's decree) they are stabilized until the electron jumps to another stationary state. The ground state($n = 1$) cannot radiate at all, so it is altogether stable against *spontaneous* decay. Of course, an electron in that state can jump upward if it is hit by a photon of appropriate energy.

The excited states($n > 1$) are all unstable against spontaneous decay. According to the principles of statistical mechanics, the atoms in a bulk sample of material at low temperature will mainly be in the ground state. Such a system will therefore display appropriate absorption lines but the emission lines will be weak.

At sufficiently high temperatures, there will be an abundance of atoms in the various excited states, and these produce emission lines as electrons decide to jump down to lower-lying levels. Notice that the frequency ω_γ of the photon emitted in a jump from n to n' is not equal to the frequency of either the parent or daughter orbital motion. But consider the case in which the jump is by one unit, from n to $n' = n - 1$. For large n, the numerator in the second factor is approximately equal to $2n$ and the denominator is approximately equal to n.

From the third of Equation it then follows that the photon frequency is approximately equal to the orbital frequency, whether that of the parent or of the daughter orbit(it does not matter which, since the two orbital frequencies are, relatively, nearly equal for large values of the principal quantum number n). This is an example of what Bohr called the *correspondence principle,* a principle that he and others exploited to guide them in the quantum thickets.

Very roughly speaking, it is the notion that in the limit where allowed orbits and their corresponding energies are closely spaced on a macroscopic scale, quantum behaviour ought to begin to resemble continuous classical behaviour.

Bohr's theory of the one-electron atom fit the data wonderfully well, though not perfectly. One correction is easily supplied. We have treated the electron as if it moves around a fixed nucleus. In fact, both the nucleus and the electron move around their common centre of gravity. This is fully taken into account simply by replacing the electron mass m in all of the above formulas by the "reduced mass" $m/(1 + m/M)$, where M is the nuclear mass and m the electron mass.

The correction is very small(for hydrogen the ratio m/M is roughly only one part in two thousand), but spectroscopic data are quite accurate enough to be sensitive to this small correction. Interest in the quantum heated up noticeably after Bohr's breakthrough, as his contemporaries sought to widen the beachhead.

How was Bohr's quantum condition to be generalized in order to deal with the noncircular orbits of the one-electron atom, the effects of external electric and magnetic fields, relativistic corrections, the vastly more complicated dynamics of many-electron atoms, and so on? Generalizations of Bohr's quantum condition suggested themselves early on to a number of people and opened the way also to considerable progress with the one-electron atom.

For example, Arnold Sommerfeld was able to treat the case of elliptic orbits in the one-electron atom. He generalized to two quantum conditions, with corresponding integer quantum numbers n_1 and n_2. He could then show that the semiminor and semimajor axes, b and a, are restricted in their relative sizes by the relation

$$b/a = n_1 =(n_1 + n_2).$$

The energy levels were however again given by Bohr's formula for the circular orbits, with

$$n = n_1 + n_2.$$

This implies a *degeneracy* in the energy levels, meaning that for a given value of n(hence of energy) there are as many different elliptical orbits as there are ways to partition the integer n into the integers n_1 and n_2. We will meet this degeneracy business again when we return to the hydrogen atom in the "modern" quantum context.

Progress with multielectron atoms was more spotty. However, the notion of discrete energy levels for atoms and molecules of whatever complexity became firmly established. It received striking confirmation from experiments involving the bombardment of atoms with beams of electrons. At low energies the electrons scatter only elastically; that is, the initial and final electron energies are the same. But at energies exceeding certain thresholds characteristic of the target atom, the electrons sometimes come off with reduced energy, the energy loss being offset by the energy gained as the atom changes its internal state.

This could be interpreted as corresponding to collisions in which the incident electron transfers energy to the atomic system, exciting it to a higher quantum level. The interpretation was confirmed by the observation that a photon of the right frequency was emitted as the atomic system then jumped back down to its initial level.

DE BROGLIE'S MATTER WAVES

A critical next step on the way to the "new" quantum theory was taken by(Prince) Louis de Broglie in the midst of his thesis work in 1923. Just as electromagnetic waves had been discovered to have particle-like aspects, he argued, perhaps ponderable matter—the electron, for example—has wavelike aspects. By a piece of luck, the following line of reasoning gave some support to this conjecture. According to Einstein, the photons constituting radiation of wavelength λ have momentum $p = 2\pi\hbar/\lambda$. Now consider an electron moving in one of Bohr's circular orbits. The magnitude p of its momentum is classically a constant of the motion for a circular orbit.

If there is some sort of wave associated with the electron, de Broglie said, it seems reasonable to suppose that the same relation between momentum and wavelength holds for the electron as for a photon. If so, it seems equally reasonable to require that the circular orbit accommodate that wavelength; namely, that the circumference be an integral multiple n of the wavelength. This leads to the relation

$$2\pi r = n\lambda = 2\pi\hbar/p;$$

hence $pr = n\hbar'$. But for circular motion pr is the orbital angular momentum L. From this chain of suppositions he thereby deduced the Bohr quantum condition $L = n'$. Einstein was impressed. He recommended approval of de Broglie's doctoral thesis.

Chapter 4

The Universe and the Gravity

The dark matter problem is one of the most important outstanding questions in cosmology today, because the precise composition and the amount of dark matter determine the ultimate fate of our Universe—whether we continue to expand, begin to contract or start to oscillate. The standard framework of modern cosmology revolves around a small set of defining parameters that need to be determined observationally in order to obtain a complete description of the underlying cosmological model of the Universe.

These three key cosmological parameters are the Hubble parameter(or Hubble constant) H_0, the mass density parameter Ω(the total matter content of the Universe, counting both the luminous and dark matter contributions) and the value of the cosmological constant. These parameters together define the physical nature and the basic geometry of the Universe we inhabit.

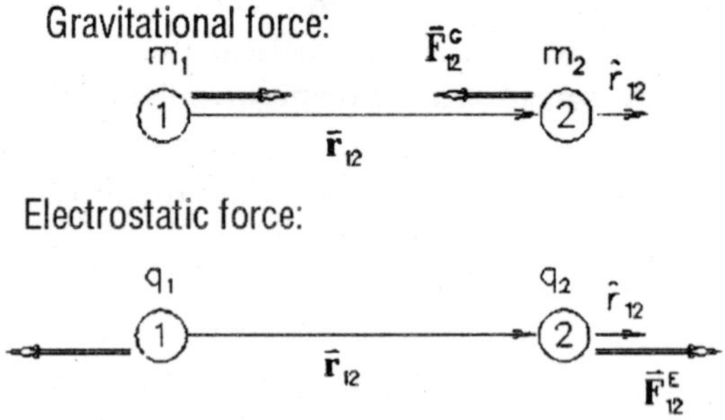

Fig. Comparison of the Gravitational force $\vec{F}_{12}^{G}$ on Mass m_2 due to Mass m_1 and the Electrostatic(Coulomb) Force $\vec{F}_{12}^{G}$ on Electric Charge q_2 due to Charge q_1.

Dark matter is defined as such since it does not emit in any part of the spectrum of electromagnetic radiation. It can therefore be probed only indirectly, principally via the gravitational force it exerts on the other masses(galaxies, stars) in its vicinity. The mass density inferred by taking into account all the visible matter in the Universe is much less than 1, therefore if W=1, as suggested by models of the inflationary Universe, then dark matter is necessarily the dominant component of the Universe and its distribution is expected to have a profound influence on the formation of all the known structures in the Universe.

The first suggestions for the existence of copious amounts of dark matter in galaxies were made in the 1920s. In 1933 Fritz Zwicky showed that there was conclusive evidence for dark matter on even larger scales, in galaxy clusters. Gravitational lensing has emerged as a powerful means of answering these questions, as it enables mass itself to be detected, as opposed to light emitted. It is an elegant technique, based on very few assumptions, and the only physics required is that of general relativity.

Lensing can, in principle, tell us about the distribution of mass in galaxies and in clusters of galaxies, and in the near future it might also provide information on still larger-scale structures in the Universe. Although it cannot directly address the question of the nature of dark matter, some lensing experiments can definitely constrain the sizes and the possible distribution of the objects that comprise it, thereby narrowing down the potential candidates.

Several dark matter candidates have been proposed, ranging from 'dark' stars—stellar-mass objects and black holes—to neutrinos, axions and many other exotic species of elementary particle. Stars which have such low masses that they are incapable of igniting the nuclear fuel in their cores, known as *brown dwarfs*, are the favoured candidates for the dark matter component in our Galaxy.

In the context of a hot Big Bang theory, neutrinos are produced in the early Universe more abundantly than baryons, so if they do turn out to possess mass, even though that mass may be very low, they can still contribute significantly to the mass density of the Universe. No cosmologically interesting limits on neutrino masses have yet been obtained, either from high-energy accelerator experiments or from the quest for solar neutrinos. Neutrinos therefore remain a viable dark matter candidate on large scales.

In our own Galaxy, evidence for the presence of dark matter comes from the observed motion of neutral hydrogen gas clouds. These clouds of un-ionised hydrogen gas follow their own orbits around the Galaxy. If they were to move only under the influence of the gravity of the visible mass, then outside the optical limits of the Galaxy their speeds ought to fall off as the square root of their distance from the galactic centre. However, these outlying clouds, detected at radio wavelengths(1400 MHz), are observed to have roughly the same orbital speed all the way out, even beyond the optically visible limit, implying the existence of an extended and invisible dark halo. The orbital motions of the globular cluster systems and the small satellite galaxies orbiting our own are also consistent with the presence of an extended dark halo that extends much farther than either the outermost stars or the limits of X-ray emissions from the hot gas that permeates our Galaxy.

GRAVITATIONAL LENSING THEORY

Gravity is one of the fundamental interactions. Because it acts at long range, it is essential to the understanding of almost all astrophysical phenomena.

Albert Einstein's theory of general relativity places the gravitational force in a physical context by relating it to the local properties of spacetime. The equivalence principle and the Einstein field equations form the core of the theory of general relativity.

The *equivalence principle* is the statement that all objects of a given mass fall freely with the same acceleration, along trajectories called *geodesics*, regardless of their composition. The curvature of spacetime —any departure from flatness—is induced by the local presence of mass.

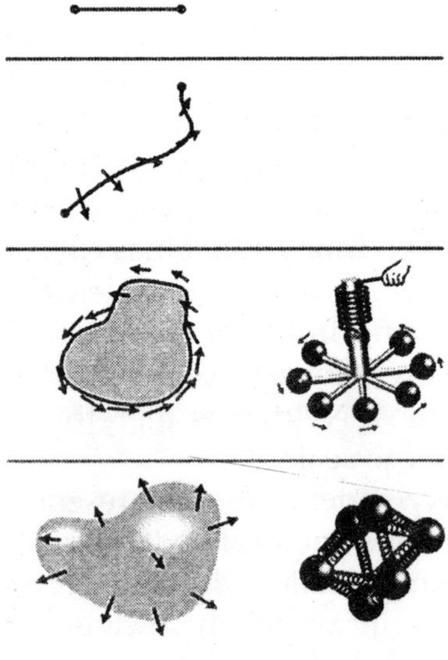

Fig. A Summary of the Derivative, Gradient, Curl, and Divergence.

The propagation of light through a lumpy Universe can easily be

understood by drawing an analogy with geometrical optics—the study of the propagation of light through media with differing densities. It is possible to make a series of simplifying assumptions to enable us to understand the lensing phenomenon. First, we assume that light propagates directly from the source to the lens, and from the lens to the observer. Second, the Universe is taken to be as described by a given mathematical prescription for the underlying spacetime, which in this case is what is called the Robertson-Walker metric.

In other words, gravity distorts the structure of spacetime. Einstein's field equations relate the curvature of spacetime called the metric to the distribution of mass and the energy content of the Universe. As a consequence, the total matter content is what determines the evolution and fate of the Universe.

The presence of mass concentrations like massive galaxies or clusters of galaxies causes light rays travelling from background sources(typically, distant galaxies or quasars) to be deflected, not unlike the bending effects caused by an optical lens. The amount of deflection produced is directly proportional to the 'strength' of the lens, which in this case is the mass, as well as to the relative orientation of the lens to the object emitting the light. Finally, the 'thickness' of the region in which the photons passing through are affected is assumed to be very small compared with the total distance they travel.

Thus, we assume that the lens can be approximated by a small perturbation to a locally flat spacetime, and also that the perturbation induced by the gravitational potential of the lensing mass along the line of sight is small.(The *gravitational potential* represents the amount of energy that the light has to expend in escaping from a concentration of mass—a *potential well.*)

The propagation of light can then be calculated in terms of an effective refractive index n, as in geometrical optics, where the path of a light ray is deflected when it crosses the boundary between two media with different properties. As in geometric optics, the propagation of light through the medium(in this case the potential) is then simply a function of the 'effective refractive index' of the lens.

We find that light slows down in a potential well, just as it slows down when it passes from one medium into a denser one. The presence of the potential well causes a deflection from a straight line in the direction of propagation by some angle, say α. This angle is given by an integral, along the path of the light ray, of the gradient of the refractive index n evaluated perpendicular to the path. Since all deflection angles are assumed to be small, the computation is along the unperturbed ray.

This now means that, for any gravitational lens, all that matters is the column density of mass of the lens enclosed within a cylinder along the line of sight. This approximation is often referred to as the *thin lens approximation*. The typical lensing geometry, where the angles θ_s, α and θ_i define the position of the source from which the light is emitted, the deflection as seen by the observer at point O, and the image position at point I. The three corresponding angular-diameter distances denoted by D_{ds}, D_d and D_s are also shown, where the subscript d refers to the deflecting lens and s to the source.

The solutions of the *lens equation*

p2000591e9950102001

help to determine the mass profile of the deflector if the image positions and relative magnifications are known.

If, for instance, θ_s=0 for all θ_i, then all rays from a source on the optic axis focus at the observer, and the appropriate lens has a uniform mass density per unit area. In most cases, multiple images of the source are seen by the observer only when the surface mass density somewhere within the lens exceeds a critical value, say Σ_{crit}. This happens typically within a small central region, whose extent is described by the *Einstein radius* θ_E.

The critical value of the mass density per unit area of the lens and the Einstein radius can be used to define an effective lensing potential on the plane of the sky. However, in most cases the source lies behind the non-critical region of the lens, in which case no multiple images are produced; instead, the images are magnified and their shapes are distorted. Since the deflection angle is proportional to the slope of the mass distribution of a lens, the scale on which only magnification and weak distortion occur is referred to as the *weak regime*.

The effect of lensing can be thought of physically as causing an expansion of the background sky and the introduction of a magnification factor in the plane of the lens. We are often interested in the magnification of a particular image, given an observed lensing geometry. Lensing conserves the surface brightness of a source along a light ray. The magnification factor of an image is therefore simply the increase in its solid angle on the sky.

For many lens models a source is significantly magnified, often by factors of 2 or larger, if it lies within the Einstein radius θ_E. An *Einstein ring,* can be formed exactly on the Einstein radius. The Einstein radius therefore marks the dividing line between sources that are likely to be multiply imaged, and those which are singly imaged.

For instance, faint circular sources that fall within the strong regime are often seen as highly elongated, magnified 'arcs', whereas small deformations of the shape into ellipses are produced in the weak regime. Therefore, looking through a lens, from the observed distortions produced in background sources(given that the distribution of their intrinsic shapes is known in a statistical sense), a map of the intervening lens can be reconstructed. This lens-inversion mapping provides a detailed mass profile of the total mass in a galaxy or a cluster of galaxies. The comparison of this mass distribution, obtained by solving the lens equation for a given configuration, with that of the observed light distribution enables constraints to be put on the amount of dark matter that is present in these systems.

At present, lensing-based galaxy mass models obtained in this fashion seem to indicate that up to 80% of the mass in a galaxy is probably dark.

DARK MATTER IN GALAXIES

When a dark mass, like a brown dwarf or a MACHO, passes in front of a background star, the light from the star is gravitationally lensed. This lensing is insufficient to create multiple images, and what is seen is simply a brightening of the background star—a phenomenon known as *microlensing*.

Since MACHOs are composed of baryons, the detection of microlensing events can help to determine how much dark matter is in the form of baryons. While the scales involved in microlensing are not large enough for multiple images to be observed, as expected in strong lensing events, the intensity of the starlight can be significantly amplified, showing up as a sharp peak in the light curve of the background star.

This was first suggested as a potentially detectable phenomenon by Bohdan Paczyñski at Princeton University in 1986. The image splitting caused by these solar-mass objects in our Galaxy is not observable, since the expected Einstein radius is measured in milli-arc seconds—well below the current resolution of optical telescopes.

Paczyñski argued that, by continuously monitoring the light curves of stars in the Large Magellanic Cloud(LMC), a satellite galaxy to our own, we would be able to observe increases in brightness that took place whenever a source in the LMC transmitted through the Einstein radius of a MACHO in our Galaxy. Since, inside the Einstein radius, magnification can occur by factors of 2 or larger, microlensing is easily detected as a sudden rise in the light intensity, independent of the observed frequency.

The probability a star being lensed by MACHOs distributed in the outskirts of our Galaxy can be estimated by modelling the lenses as point masses. The quantity needed to compute the number of expected events is referred to as the *optical depth to lensing*, which is simply the chance that a given star in the LMC lies within the Einstein radius of a lens at a given time. The optical depth is calculated along the line of sight, and it depends on the total assumed number density of MACHO lenses.

There are currently several observational research groups searching for microlensing signatures in LMC stars and stars in the galactic bulge by continuously monitoring the light curves of millions of stars. Looking towards the centre of our Galaxy, we seek to detect MACHOs in the disk, and looking in the direction of the LMC we seek MACHOs distributed in the galactic halo.

Several large international collaborations, known as MACHO, EROS, DUO and OGLE, are currently engaged in this venture. When the MACHO group analysed the data from their first-year run, consisting of almost 10 million light curves, they detected one event with significant amplitude in the magnification, and two with modest magnifications.

They estimated the total mass of MACHOs inside a radius of 50 kiloparsecs to be around 8×10 solar masses. This result was found to be reliable and fairly independent of the assumed details for the underlying halo model. However, it is clear that the fractional contribution to the halo mass from these MACHOs is small. For instance, within the mass range of 3 × 10" to 6 × 10" solar masses,

MACHOs account for significantly less than 50% of the halo. At the end of their second year of accumulating data, now with six to eight events, they estimated a halo fraction of 30% to 90% in the mass range 0.1 to 0.4 solar masses.

The picture that emerges of our Galaxy in the light of the results from these microlensing searches is that, perhaps, a significant fraction of the dark matter content of our halo is baryonic, and is distributed in stellar-mass objects.

Lensing by a galaxy, with a typical mass of 10 solar masses, instead of by a star of 1 solar mass, produces splittings of an arc second or so between the multiple images. The first lensing galaxy, designated 0957 + 561A, was discovered in 1979, and as of early 1998 more than 30 such gravitational lenses were known.

Since the lens magnifies a faint background galaxy or quasar, it acts as a gravitational telescope and enables us to see farther than we can ever probe using either ground-based telescopes or instruments in space. For multiple image configurations, since the different light-ray paths that correspond to the different images have different lengths, relative time delays can be measured if the source is variable. A successfully 'inverted' lens model can be used to measure the Hubble constant H_0, the precise value of which has implications for both the age and the size of the Universe. H_0 can be determined from lensing, in theory, by measuring two quantities: the angular separation between two multiple images, and the time delay between those images.

If the source itself is variable, then the difference in the light travel time for the two images comes from two separate effects: the first is the delay caused by the differences in the path length traversed by the two light rays from the source, known as the *geometric time-delay,* and the second is a general relativistic effect—the *gravitational time-delay*—that causes a change in the rate at which clocks tick as they are transported through a gravitational field.

And since the two light rays travel through different portions of the potential well created by the deflecting lens, the clocks carrying the source's signal will no longer be synchronised when they emerge from the potential. Once these time delays, the image separations and their relative magnifications are measured, the distance to the lens and the source can be deduced from the lens equation, which then allows an independent estimate of H_0 to be made. Quasars are ideal subjects for lensing since they are very luminous, lie at cosmological distances and hence have a high lensing probability. The first multiply imaged quasar, QSO 0957 + 561A, B, was discovered in 1979 by Walsh, Carswell and Weymann. The lensing of this distant quasar at a redshift of $z = 1.41$ is caused by a bright elliptical cluster galaxy at $z = 0.34$. This system has been continuously monitored for several years, since it was thought to be an ideal candidate for estimating H_0 from the measured time-delay.

Detailed modelling has provided estimates of the properties of the lensing galaxy(such as its mass and density profile) which are in good agreement with the values obtained from independent dynamical studies. For the 0975 + 561 system, there has been some disagreement between different groups that have attempted to measure the time-delay from the offsets of the light curves of the two images, leading to two estimates of the Hubble constant that differ by 20%.

At present there are several systematic surveys under way aimed at detecting both large and small multiple-imaging lenses in the optical and radio wavebands. Therefore, while lensing is at present unable to provide a precise measurement of the Hubble constant on the basis of the candidate multiple image systems detected so far, perhaps the ideal 'golden lens' is waiting to be discovered.

Massive foreground galaxies can also lens fainter background galaxies, and this effect can be used to examine several interesting issues. The frequency of galaxy-galaxy lensing provides a glimpse into the redshift distribution of galaxies, and the distribution of mass at high redshifts, and gives us an idea of typical mass distributions in galaxies. Galaxy-galaxy lensing is expected to produce mainly weak effects, such as an apparent increase in the statistically small likelihood of a ring of faint background galaxies occurring around bright foreground galaxies.

A tentative detection of such a signal has been reported, and the results seem to indicate that isolated galaxies have very large dark halos extending out to around a hundred kiloparsecs from their centres. Dynamical estimates of the mass distribution of isolated, non-cluster galaxies obtained by mapping the motion of satellite galaxies in orbit around them also seem to indicate that, while luminous matter dominates in the inner regions of galaxies, in the outer regions dark matter can constitute up to 90% of the total mass.

DARK MATTER IN CLUSTERS OF GALAXIES, AND BEYOND

Clusters of galaxies are the most recently assembled and largest structures in the Universe. Clusters are more complex systems and harder to understand than stars, for instance, since their formation necessarily depends on the initial cosmic conditions.

A typical rich cluster contains roughly a thousand galaxies, plus gravitationally bound, hot, X-ray emitting gas; and there is strong evidence for the presence of significant amounts of dark matter(comprising about 90% of the total mass of the cluster).

The currently accepted theories for structure formation in a Universe dominated by cold dark matter postulate that dark haloes essentially seed the formation of visible galaxies. Cosmic structures are also expected to build up hierarchically, small objects forming first and then aggregating together, driven primarily by gravity, to form larger units. In the standard picture, each galaxy forms within a dark halo as a result of the gas collapsing, cooling and fragmenting to form stars. It is believed that when galaxies, along with their dark haloes, hurtle together to form a cluster, the individual haloes merge into a large, cluster-scale dark halo.

Lensing of background galaxies by clusters can be divided into strong lensing, in which giant arcs are observed, and weak lensing, in which images of background galaxies are weakly distorted, producing 'arclets'. For a general lens model the number of images obtained from a compact source is odd: one image is obtained if the source is far away, but as the distance decreases it crosses curves known as *caustics*.

Every time a caustic is crossed, the number of images increases by two. Giant arcs are observed because the magnification of a source is greatest when it lies on a caustic. Giant arcs may be used to investigate the mass distribution in clusters, in much the same way that the lens model inversion method can reveal the mass distribution in galaxies. There are now several successfully modelled lensing clusters, where the mass maps obtained agree well with those determined from the clusters' X-ray emission and by applying the virial theorem to the motions of cluster galaxies.

STRONG AND WEAK LENSING

For weak lensing by an extended lens, and in the thin-lens approximation, ray-tracing methods borrowed from geometric optics may be used to map objects from the source plane into the image plane in the process of solving the lensing equation. Several properties of lensing can be used to refine this mapping:

- The conservation of surface brightness, as in conventional optics;
- The achromatic nature of lensing(i.e. lensing effects are independent of the frequency of the light emitted by the source);
- The fact that the deflection angle does not vary linearly with the impact parameter.

Lensing produces two distinct physical effects: the convergence or magnification(κ) is the focusing term that represents simple magnification produced by matter enclosed within the beam; $\kappa > 1$ corresponds to strong lensing, which gives rise to multiple images and arcs.

The second effect is the *shear*(γ), which is the anisotropic distortion of images that lie outside the beam produced by the gradient of the potential; $\kappa \cong 0$ and $\tilde{a} > 0$ corresponds to weak lensing, which gives rise to distorted images(arclets) of the faint background sources. The total amplification is a sum of the contributions from both these effects.

Strong lensing is observed in the multiply imaged region where the surface mass density, Σ, exceeds Σ_{crit}. The number of multiple images is determined by the precise configuration, the redshift distribution of the sources(which is in general unknown) and an underlying cosmological model.

Giant arcs have been observed around some 30 clusters, primarily by the exquisite imaging capabilities of the Hubble Space Telescope(HST). Giant arcs, which are typically images of spiral galaxies at high redshift, are defined as having an axis ratio(the ratio of the long axis to the short axis) in excess of 10.

The curvature of the arc is a measure of the compactness of the mass distribution of the lensing cluster, since the radius of the arc corresponds roughly to the Einstein radius.

The rotation curves along the length of arcs have been mapped for the Abell(dense) clusters Abell 2390 and CL 0024 and found to be flat, indicative of the presence of a dark halo. In principle, if the true luminosity of the lensed galaxy is known, this technique can be used to extend the extragalactic distance scale to objects with very high redshift. Detailed modelling of cluster cores requires the following ingredients: arc positions, the number of merging images and whether this number is odd or even, arc widths, shapes and curvature to constrain the location of critical lines on the image plane.

Given one or more measured redshifts of the arcs, the mass enclosed within the arc can then be accurately estimated, enabling the lens model to be refined. Many cluster cores have been successfully studied from their strong lensing features: Abell 370, Abell 2218, AC 114 and MS 2137–223 to name a few.

The HST's imaging power uniquely helps in the identification of multiple images, so these models can be used to assess the smoothness of the dark matter distribution. The results obtained with these models demonstrate that the total mass distribution in a cluster closely follows the luminous mass.

The overall ratio of the total mass to the total light measured in the visual band in solar units(i.e. in terms of the Sun's mass and luminosity) ranges from 100 to 300, in good agreement with the values of 150 to 250 obtained by independent methods.

In weak lensing by clusters, single images are obtained, but they are sheared as well as magnified. The deformation in shape produced by the lens can be related directly to the contributions from the deflecting mass if the shape of the source is known, but unfortunately this is rarely the case.

We therefore have to proceed by statistical methods, assuming that there is a distribution of shapes. An elegant 'inversion procedure' can be used to obtain a map of the mass density in the plane of the lens from the statistics of these sheared shapes. This map is only relative, since a uniform sheet of dark matter will produce no detectable shear. The mass density obtained by this method is therefore only a lower limit to the true mass: if a uniform sheet of material were added, the observed results would not change. Several variants and refinements of this basic scheme have been developed and successfully applied. The total amount of matter that is suggested by these measurements is such that the mass-to-light ratio typically lies in the range 200–800 solar units.

These values are consistent with estimates obtained on comparable scales from X-ray observations. Since the mass-to-light ratio measured for the luminous parts of galaxies ranges from 1 to 10 solar units, indicating that large amounts of dark matter must be present in clusters, as first proposed by Fritz Zwicky. While most inferred total mass distributions roughly follow the distributions of luminous matter, some clusters seem to have a more centrally concentrated mass distribution than is traced by the galaxies, while others have mass distributions that are much smoother the than the light distribution.

Aside from providing mass estimates for individual clusters independently of any assumptions made about their dynamical state, the ultimate goal is to determine the relative numbers of clusters of different masses, since that is a strong test of the underlying cosmological model. Some recent research has focused on combining the information obtained for a cluster in the strong and weak lensing regimes to build composite mass models. One question that has been tackled is that, if all individual galaxies have massive and extended dark haloes, then what is the fate of these haloes when the galaxies hurtle together to form a cluster? What fraction of the dark matter gets stripped and redistributed?

By applying lensing techniques to a very deep, wide-field HST image of the cluster AC114, it is found that 0.1 average a bright cluster galaxy has only two-thirds the mass of a comparable non-cluster counterpart, indicative of mass-stripping having occurred. The halo size is also much more compact than that of an isolated galaxy. The conclusion at present is that only 10% to 15% of the total mass of a cluster is associated with the member galaxies, and the rest is probably distributed smoothly throughout the cluster.

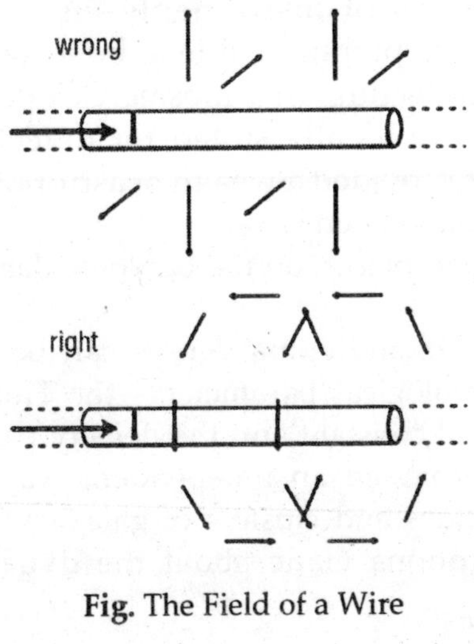

Fig. The Field of a Wire

Since gravitational lensing is sensitive to the total mass enclosed within a cylinder along the line of sight, we can potentially reconstruct the power spectrum of mass fluctuations that over time have been amplified by gravity, leading to the formation of massive large-scale structures.

In the standard scenario, very massive objects like superclusters and filaments are expected to form, and they can be probed by the weak lensing signal they induce in background galaxies.

In this case it is not the surface mass density that is reconstructed, as with clusters, but rather the power spectrum of density fluctuations. The distortions that are measured can be related to the fluctuations of the gravitational potential along the line of sight. At present, there have been no unambiguous detections of shear on scales larger than clusters, but the prospects are encouraging.

Great strides have been made in probing dark matter using gravitational lensing to map the mass distributions of galaxies and clusters of galaxies.

Theoretical progress in the future is expected primarily in the field of improved mass-map reconstruction techniques and their applications to probe the mass distribution in galaxies, clusters and other large-scale structures.

Extending existing methods to detect coherent weak shear induced by still larger-scale structures like filaments and superclusters is the next step. In order to make any further observational headway in the detection of weak shear induced by the intervening large-scale structure, we need wide-field images that probe down to much fainter magnitudes.

The new generation of instruments—including the Hubble Advanced Camera for Exploration, due to be installed on the HST in 1999, and the large-collecting-area mosaic CCD detectors currently under construction—are ideally suited for detecting shear to high precision. Lensing has provided a wealth of astrophysical applications. The most significant have been:

- Limits have been placed on the baryonic dark matter content of our Galaxy;
- The properties of individual lenses can be used to refine the values of cosmological parameters—the Hubble constant H 0, the cosmological constant and the density parameter Ω;
- Lensing has provided an independent way of measuring the masses of galaxies and clusters of galaxies that is independent of any assumptions made about the dynamical state of the system;

- It simulates a giant gravitational telescope that offers a view of the distant Universe that would otherwise remain inaccessible.

It has provided essential clues to the evolution of galaxies by enabling the mass profiles(inferred from lensing) in dense environments like cluster cores to be compared with those of isolated, non-cluster galaxies.

Chapter 5

Faraday's Unified Field Concept

Strongly influenced by the 18th century thinker, Roger Boscovich, Michael Faraday, in the 19th century, introduced the field concept to physics with its application to the description and understanding of the phenomena of electricity and magnetism.

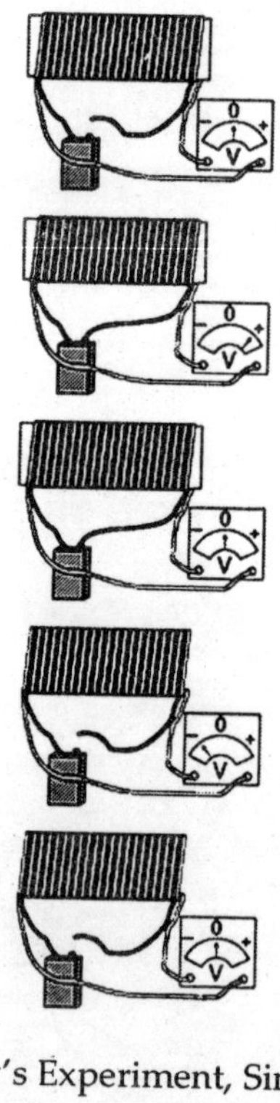

Fig. Faraday's Experiment, Simplified and Shown with Modern Equipment.

He speculated that, contrary to the atomistic view of matter, its fundamental description should instead be in terms of continuous fields of force. These were to represent the strength with which matter affects matter, at the continuum of points in space and time. Calling this description 'fundamental' meant that it was the continuous field of force that was to be taken as the essence of matter, rather than considering this as a secondary(derivative) feature of a quantity of matter.

The 'thing' feature of the atomistic view, for example in terms of its position, momentum and energy, all localized in space, was then taken to be secondary-to be derived later, as a consequence of the primary field of force-rather than the other way around.

To exemplify this conceptual difference between Faraday's continuum view and Newton's atomistic view, consider the interpretation of the(empirically confirmed) law of universal gravitation. If any two quantities of matter, with inertial masses equal to *m*1 and *m*2, should be separated in space by the distance *R*12, then the force of attraction between them is

$$F = G\frac{m_1 m_2}{R_{12}^2}$$

Faraday interpreted this formula in a non-atomistic way as follows. Consider a potential field of force that could act on a test mass somewhere, if it should be there, as the continuous field,

$$P_2 = G\frac{m_2}{R^2}$$

where *R* is a continuously varying spatial point, measured from some appropriate spatial origin, for the problem at hand. The field of force, *P*, is then defined continuously everywhere, except at the origin, *R* = 0. If we now consider a test mass, *m*1, to be placed at *R* = *R*12, then the observed force at this place would be the coupling of this test mass to the field of force *P*2 at this point of observation. That is, the force on the test body at the location *R* = *R*12 is the coupling of this mass to the field:

$$m_1 P_2 = m_1 \frac{Gm_2}{R_{12}^2}$$

in agreement with the empirical expression for the gravitational force. The constant *G* is called 'Newton's gravitational constant'-it is the same for all gravitational interactions.

Faraday then asserted that the continuum of lines of force, defined by the field *P* for gravitation(or by other fields for other types of force, such as the cause of the magnetic alignment of a compass needle or the cause for the electrical repulsion between two bits of amber that had been rubbed with fur), are the fundamental stuff from which any theory of matter should be built.

Faraday(and Oersted before him) did not believe that there are fundamentally different types of forces between matter and matter. He felt, rather, that all forces must be unified into some sort of universal force-an entity that would only appear as disconnected types of force under different sorts of experimental circumstances.

In Denmark, Hans Christian Oersted first discovered that a compass needle will become aligned in a particular way in a plane that is perpendicular to the direction of flow of an electric current. Faraday then interpreted the magnetic field of force(in this case, acting on the compass needle) as no more than a representation of an electric field of force in motion, that is to say, an electric force as observed in a moving frame of reference relative to a stationary observer(that of the compass needle). Since 'motion', *per se,* is a purely subjective entity, the implication then followed that the electric and the magnetic fields of force are actually no more than particular manifestations of a unified electromagnetic field of force.

It also implied, because of the relativity of motion, that an electric field is nothing more than a magnetic field in motion! [This idea led to the invention of the dynamo-so important to the industrial revolution of the 19th century.]

The relation between polarization, magnetization, bound charge, and bound current is as follows:

$$\rho_b = -\Delta \cdot P$$

$$J_b = \Delta \times M + \frac{\partial P}{\partial t}$$

$$D = \epsilon_0 E + P$$

$$B = \mu_0 (H + M)$$

$$\rho = \rho_b + \rho_f$$

$$J = J_b + J_f$$

where P and M are polarization and magnetization, and ρ_b and J_b are bound charge and current, respectively. Plugging in these relations, it can be easily demonstrated that the two formulations of Maxwell's equations given in Section 1 are precisely equivalent.

As we have discussed earlier, James Clerk Maxwell, in expressing Faraday's unified field theory in a rigorous mathematical form, showed further that such unification also incorporates all of the known optical phenomena-the ray properties of light(e.g. the focusing of light through lenses, etc.) as well as the wave features-its polarization characteristic(being polarized in a plane that is transverse to its direction of propagation,(as originally discovered by Augustine Jean Fresnel), its properties in combination with other optical waves yielding interference and diffraction, etc. The Maxwell field theory also predicted other radiation phenomena, such as radio waves, X-rays and gamma rays.

With his discovery of the unification of electricity and magnetism, together with optics, Faraday believed that this was only a first step toward a fully unified field theory of force-one that would incorporate the gravitational force as well.

Ingenious as he was, Faraday was not able to devise any experimental scheme that would verify such a unification of electromagnetism with gravity. For example, he was not able to convert electric or magnetic currents into a gravitational current, or to demonstrate the existence of gravitational radiation that could fuse with electromagnetic radiation.

From the theoretical side, there were also some unanswered questions on the validity of such a unification. An important one concerns the reason why, on the one hand, electromagnetic forces can be either attractive or repulsive, while, on the other hand, gravitational forces have only been found to be attractive.

In the century that preceded that of Faraday, Boscovich was concerned with this question. He attempted to answer it by describing gravitational forces in terms of a continuous effect that matter has on matter, as a function of their mutual separation, in such a way that the attractive character of the force changes smoothly into a repulsive force when sufficiently small mutual separations are reached.

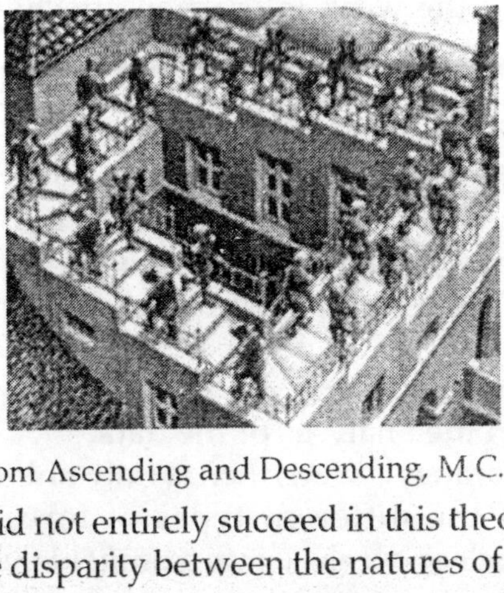

Fig. Detail from Ascending and Descending, M.C. Escher, 1960.

However, he did not entirely succeed in this theoretical task, nor did Faraday explain the disparity between the natures of the electromagnetic and the gravitational forces, nor did Einstein resolve the problem in the contemporary period. But all of these investigators did believe that the approach of a fully unified field theory that would incorporate the electromagnetic and the gravitational forces, along with any other type of force that may exist in nature, such as the nuclear(strong and weak) interactions-that were not known yet in the 18th and 19th centuries-was indeed in the direction toward a general theory of matter.

The aim was to construct a unified field theory that would incorporate the electromagnetic, gravitational and any other type of force to be discovered in the future-in a way that the universal force field would manifest itself as different sorts of force under correspondingly different types of experimental conditions, such as the manifestation of the magnetic force when the viewer is in motion relative to the reference frame of an electrically charged body.

In the 20th century, the other types of force that revealed themselves were in the domain of atomic and nuclear physics. The nuclear(strong) force, that binds neutrons and protons in an atomic nucleus, is the order of a hundred thousand times stronger than the repulsive force between two protons due to their mutually repulsive electric force.

In this domain, the electric force between the protons is the order of 10 times stronger than a newtonian force of gravity between them. Also in the nuclear domain there are the 'weak interactions', which are responsible, for example, for beta decay of radioactive nuclei-the conversion of neutrons(or protons) in nuclei into protons(neutrons), electrons(positrons) and neutrinos.

In this domain, the strength of the weak interaction is around a hundred times weaker than the electromagnetic interaction. Both the nuclear(strong) and the weak interaction effectively 'turn off' as the mutual separation exceeds the size of a nucleus, around 10" cm. On the other hand, the electromagnetic and the gravitational interactions have infinite range-that is, they never really 'turn off'.

In contrast with the electromagnetic and gravitational forces, the nuclear force has a part that is attractive and a part that is repulsive. Which one dominates depends on the conditions imposed, while the other is masked. Different types of nuclear experimentation, mostly involving the scattering of nuclear particles from each other, are able to project out the repulsive and attractive components of the nuclear force, separately, in the analyses of the data.

According to Faraday's view, then, a fully unified field theory must incorporate all of the different types of force-the gravitational, electromagnetic, nuclear, weak, ...into a unifield field of potential force, with each generic type of force manifesting itself with regard to the behaviour of a test body, under correspondingly different types of experimental conditions.

Finally, it is important to note once again that 20th century physics has revealed the fact that the inertial features of matter are intimately related to what seems to be an entirely different sort of continuous field-the field variable that has been interpreted as a 'probability amplitude' in quantum mechanics.

A fully unified field theory would then also have to incorporate the successful formulae and equations of quantum mechanics in order to describe the atomic domain satisfactorily. The formalism of quantum mechanics that is used today may then appear as an approximation within a more general field description that would unify the force manifestations of matter(gravity, electromagnetism, nuclear, ...) and the inertial manifestations of matter(the feature of resisting a change in its state of constant motion, due to imposed forces).

That is to say, a fully unified field theory must incorporate the actions of bodies on each other(the forces) and their respective reactions(their inertia). Question Is there any logical reason, that is, aside from aesthetics, that all of the laws of nature must be derivable from a single, general theory? Reply It is true, of course, that 'beauty is in the eye of the beholder', but in addition to the reaction that I have(or that you may have) to the idea of a complete, unified scheme for all of the different types of physical phenomena, as being more beautiful than a patchwork of disconnected theories, there is also an important logical reason for striving toward a general theory. A patchwork scheme runs into difficulties on the question of logical consistency, especially at the 'seams' where the patches are supposed to be joined.

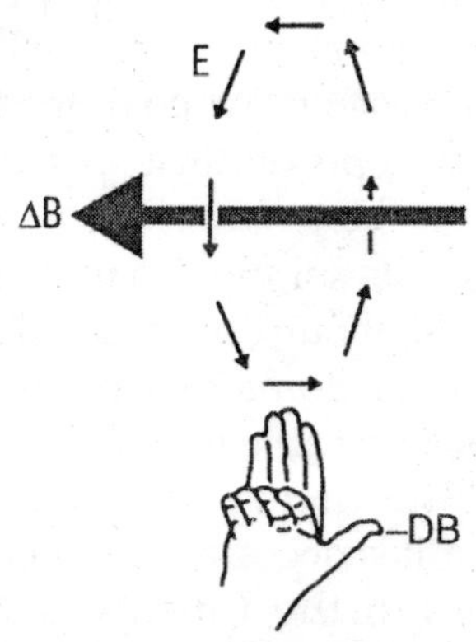

Fig. The Relationship Between the Change in the Magnetic Field, and the Electric Field it Produces.

Consider, for simplicity, two physical theories, initially constructed to predict two seemingly mutually exclusive sets of phenomena. Suppose that these theories are such that a part of the axiomatic basis of one of them is logically incompatible with some of the underlying axioms of the other.

For example, one theory might assert, axiomatically, that one quantity of matter exerts a force on another only when they are in contact. The second theory may assert, in contrast, that forces between distant quantities of matter are exerted spontaneously, over any distance. Now it may happen that as the methods of experimentation improve, the conditions which dictate the criteria for using one of these theories or the other will fuse into each other-then requiring the use of both theories at once!

A well-known example is the theory of general relativity, replacing Newton's theory of universal gravitation. These are certainly logically dichotomous theories; if taken together they would be a logically inconsistent, single theory.

Einstein's theory is based on the continuous field concept and the prediction of a finite speed of propagation of interactions between the material components of the system; Newton's theory is based on the discrete atomistic model of matter, and the notion of spontaneous 'action-at-a-distance'.

If one should allow the speed of one gravitationally interacting body to be continuously slowed down, relative to another, and their separation is allowed to become sufficiently small, so that in a minute portion of space the curved space-time in their domain may be replaced with a flat space-time(the tangent plane to the point of observation in a curved space), then, Einstein's field equations can be approximated by Newton's equation for the gravitational potential.

Trying simply to adjoin Einstein's theory to that of Newton then leads to the question of logical consistency when trying to decide precisely where should one 'turn off the concepts of Einstein's theory and 'turn on' those of Newton's theory.

There is no unambiguous solution to this problem! It is then for the sake of logical consistency that one cannot consider both of these theories as correct, even in the proper domains where they 'work'. To maintain logical consistency-that is, to be able to say that particular empirical data can be 'explained' by one theory and not another theory that is incompatible with the first-it is necessary to foresake one of these theories for the other, even though the foresaken theory mathematically approximates the accepted theory, under those experimental conditions where the foresaken theory 'works'. It is, in fact, the latter comparison that gives us hints about the character of a more general theory-through the requirement that this formalism must approach the form of the earlier theory, in the limit where the latter theory was empirically successful.

This is known in the history of science as a 'principle of correspondence'. It has been used throughout the different periods of what Kuhn has called 'normal science', and their transition through a scientific revolution, to the next period of normal science. It is in this sense that continuity persists throughout scientific revolutions-even though many of the ideas of the superseded theories are abandoned. Question: Is there any comparison between the evolution of ideas in science and the evolution of the human society?

Reply: I think that the continuity of ideas in science is analogous to a continuity that persists in certain features of the evolution of the human society, in proceeding from one generation to the next. Certainly all of the concepts adhered to by an older generation are not maintained in all succeeding generations!

But there are notions, values and facts that do persist. That which remains had usually been sifted, refined, re-evaluated, and taken in a new perspective.

I would then contend that without these concepts and attitudes that had been passed on from one generation to the next in this way, there could have been no progress at all in our comprehension of the world, nor could there have been a positive propagation of values of a civilized society. Little as this progress has been since the infancy of the human race, I believe that it is nevertheless nonzero, as exemplified in the laws of society that we do have to protect the rights of all people.

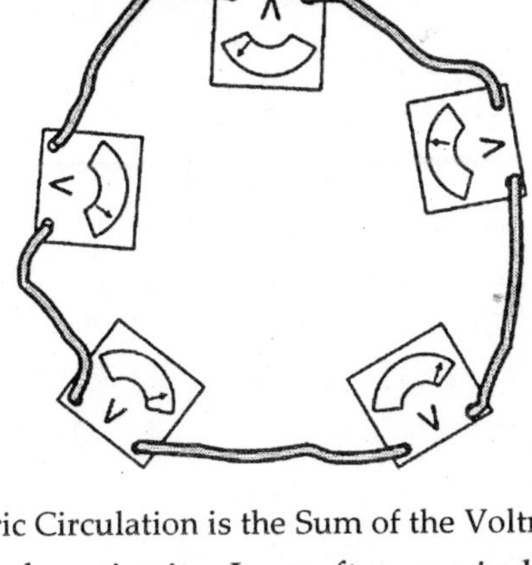

Fig. The Electric Circulation is the Sum of the Voltmeter Readings.

On the notion of continuity, I am often reminded of a conversation I was told about between a former colleague of mine and his nine year old daughter, Suzie. One morning, at breakfast, Suzie asked her Dad,

'Where was I before I was born?' He replied, 'You were nowhere before you were born! You didn't exist!' Without any hesitation, Suzie replied, 'I don't believe it. It can believe that you may not know where I was before I was born, but I must have been somewhere!'

Suzie's Dad then took a blank napkin from the table and a pen from his pocket.

He showed her the blank napkin and asked, 'What do you see on this napkin?' Suzie answered, 'Nothing'. Then her Dad drew a circle on the napkin with his pen and asked: 'What do you see on the napkin now?' Suzie answered, 'A circle'. Convinced of his argument, Suzie's Dad then asked her, 'Where was the circle before I drew it here on the napkin?' Suzie replied, again without hesitation, 'It was in the pen!'

I suspect, and I related this to my colleague when he told me about the conversation, that, perhaps because of her beautiful lack of inhibitions, Suzie had a more subtle, profound and scientifically valid answer to her own question than her Dad did!

For the next question would have been: 'Where was the circle before it got into the pen?' This would have led to Suzie's father's mind, leading in turn to his experience with the world, ... and so on, *ad infinitum,* until the entire universe would have been used up! This final answer may then have viewed the world holistically, in terms of all of its seemingly different manifestations(people, planets, galaxies, ...) from the approach of a unified field theory!

Chapter 6

Einstein's Unified Field Concept

During his later years, Einstein attempted to show that, at least the electromagnetic field of force could be unified with the gravitational field of force, in terms of a more general sort of geometry than that which he found earlier to represent accurately the gravitational force alone. For if the gravitational force could be derived as a feature of the motion of a freely moving body in a space-time governed by the relations of a riemannian geometry, and if, in principle, there is nothing exclusive about the gravitational force, as compared with any other type of force in nature, then a generalization of the geometry was a natural step toward further extension of the theory of general relativity, which Einstein considered to be its completion.

After all, in addition to their inertial properties, gravitating bodies do possess electric charge, so that when brought sufficiently close together their mutual electromagnetic interaction does 'turn on', dominating the gravitational force in this region.

But even when it does dominate, the gravitational force is present-they are there together, though one of them is masked. That is, in the limiting process, it seems unreasonable that at some(unprecisely defined) juncture, the riemannian geometry would suddenly 'turn off' and the euclidean geometry would simultaneously 'turn on', along with the action of the electromagnetic forces.

It seems more reasonable, at least from Einstein's point of view, that there should exist a generalization of the geometry of space-time that would, in principle, unify the gravitational and the electromagnetic fields, under all possible conditions-giving an apparent view of pure electromagnetic forces at sufficiently small separations and an apparent view of pure gravitational forces at sufficiently large separations-but both actually being present at all times, with one or the other dominating.

In addition to the role of geometry in Einstein's unified field theory(a feature not present in Faraday's unified field approach), there is also the following important conceptual change. In Faraday, the field of potential force represents matter-it is its essence. The test body is then brought into the picture to probe this field. In Einstein's field theory, on the other hand, especially when it incorporates the Mach principle, the test particle is not, in principle, separate from the field of force that it is supposed to probe. Rather, there is one closed system to consider, with its underlying continuous field that incorporates, in a unified way, all interacting components.

The test body is then abstracted from the description of this closed system in an asymptotic limit, where the particular component of the whole system only appears as a weakly coupled 'part'. This abstraction is carried out by taking a special approximation for the full mathematical expression of the closed system.

The two views are conceptually inequivalent. Einstein's unified field theory approach is a generalization of Faraday's in the sense of predicting, in principle, more physical features than are predicted by Faraday's approach.

One reason is that in Faraday's field theory the test body is ambiguously defined; for example, it may have an electric charge if it is to probe the strength of an electric field of potential force, but none other of its fundamental properties are involved in the basic theory.

Fig. A Ground Fault Interrupter.

In Einstein's theory, though, the test body is fully and uniquely defined in all of its characteristics-following from the asymptotic features of the entire closed system that one starts with. Thus, while more of the features of the physical system are implicit in Einstein's theory, all of the predictions from Faraday's theory are also contained in the predictions of Einstein's theory.

Thus, Einstein's approach to a unified field theory is a true generalization of Faraday's approach to a unified field theory. A major conceptual change from Faraday's unified field concept to that of

Einstein, regarding what has just been said about the different ways in which the test body is introduced, is the difference between the 'open' system, that characterizes Faraday's theory, and the 'closed' system, that characterizes Einstein's theory.

For example, Faraday's field of force is a linear superposition of fields-a(vectorial) sum of the force fields for each of the component 'influencers' in the system, that could act on a test body. Here, the test body is just one more object-whose field is summarized in terms of a set of particle-like qualities associated with its localization-such as charge, mass, magnetic moment, etc.

Thus, Faraday is still talking about many individual 'things', with his field theory, although they are most fundamentally characterized by their separate fields of influence-later to be added in order to determine the total effect on a test body.

Einstein's unified field theory, when it is expressed in a way that is compatible with its meaning, and when it incorporates the Mach principle, is to represent only one thing-this is the single, closed system, not composed of separate parts.

This is the universe. It is a philosophy that implies holism. The apparent view of a system in terms of separable parts is, here, only illusory. It appears that one of the inseparable components of the closed system is indeed a 'thing' that is weakly coupled to the rest of the system-like the coupling of our planet, Earth, to its entire host galaxy, Milky Way-so much so that we could extrapolate to the limit where they are actually uncoupled things! But the latter extrapolation is, in principle, impossible, according to Einstein's conceptual stand.

This is because the test body is still only a particular manifestation of the entire closed system, just as a ripple of a pond is only a particular manifestation of the entire pond, rather than being a thing-in-itself. One cannot remove a ripple from the pond and study it on its own, as an independent 'thing', with weight, size, and so on.

Nevertheless, from the mathematical point of view, one can, with accuracy, assume that the weakly coupled entity(like the ripple) is practically a disconnected entity. One may then derive its motion as the dynamics of an independent body, from the equations of motion that would describe such an independent thing, perturbed by the rest of its environment.

It is still important, however, that one must view the latter mathematical description as no more than an approximation for an exact description of the entire closed system-a system that is in fact without independent parts.

The latter more general mathematical structure is quite different than the structure of the equations that describe the approximated situation. These differences might show up in other predictions, even when the system looks like it has an independent part in it, though weakly coupled to the rest.

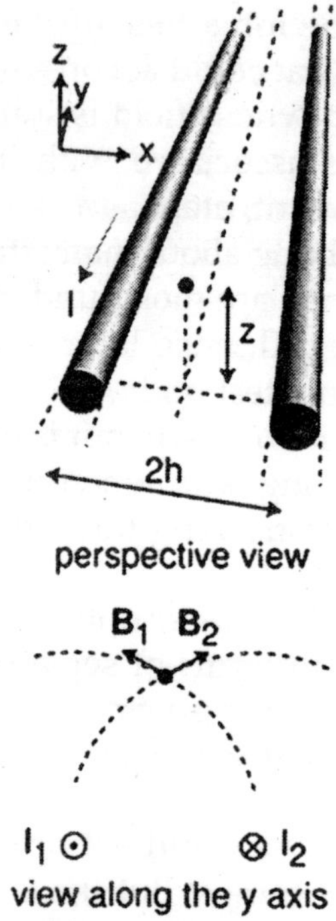

Fig. Vector Addition

Question How does the philosophy that underlies Einstein's unified field concept compare with earlier views of closed systems, such as those of Spinoza and the monad theory of Leibniz?

Reply The philosophy that follows from Einstein's unified field concept, attempting to express all physical manifestations of matter in a single conceptual scheme, does indeed bear some resemblance to the Leibnizian world as fundamentally one, without separable parts-from my understandii.g of his philosophy.

Plato also had the view that the universe is a single, existential entity, though he believed that its form is changeable.

Leibniz viewed the world in terms of what he called 'monads'-entities that many philosophers have likened to atoms. But Leibniz' monads were not really separable things, as are the constituent elements of the real, material world, according to the atomists of ancient Greece, or the atomists of the contemporary period-such as Bohr and

Heisenberg. Leibniz' monads are different from separable atoms-they are more like the ripples of a pond, or the notes sounded by a violin string-as particular modes of behaviour of a closed system. Leibniz referred to his assemblage of monads as an array of reflections of one world, I would then interpret his word 'reflection' to mean 'physical manifestation'.

With this interpretation, Leibniz would not be taking the atomistic view at all; rather, he would be taking an approach more similar to that of Einstein, where it is assumed that there is only one, real, objective world, without actual parts, but rather characterized by an infinite spectrum of possible physical manifestations(monads).

It is interesting to extrapolate from this holistic view of the material world to a universe that includes humankind. Assuming, with Spinoza, that such an extension is logically necessary, one might then view the human beings' inquiry into the nature of the universe(including themselves!) as based on an approximation that our consciousness can get into, thereby reflecting on underlying abstract relations about the nature of the real world.

We may then extrapolate from these relations toward a more complete understanding of the universe-that fundamental existent of which the human being is, in principle, an inseparable component. According to this view, the human being's attempt to comprehend the world is not a matter of our 'looking in', as impartial observers, tabulating its physical characteristics, according to the responses of our sense impressions.

It seems to me to be, rather, a matter of the human being's intellectual reflections, introspections and deductions about the basic nature of a single, abstract, underlying reality-that existent that is the universe, and from which the human being derives particulars to be correlated with the physical reactions of our senses.

Still, we must keep in mind the idea that there are infinitely more particulars that fall within the scope of our reasoning power, but are not within the domain of direct responses to our sensing apparatuses. Nevertheless, the latter(unobservable) particulars can play the important role of the logically necessary ingredients of a theory, that leads, by logical implications, to bona fide tests of our comprehension that do entail directly observable particulars.

There are critics of such a unified approach who argue that, according to Gödel's theorem, in logic, it is illogical to start at the outset with a complete system, because the investigators themselves must have the freedom to decide how, where and when to explore along one direction or another, i.e. they have the freedom to exercise a decision-making process, with their own 'free wills'. But this view tacitly assumes that the human being is indeed a separable entity, apart from the rest of society and everything else in our environments! Should one accept this assumption as a fundamental axiom, then Gödel's theorem could apply to the real world, outside of the logical system to which he applied it(a theory of arithmetic).

One would then not be able to assert that there could be 'complete knowledge' to talk about in the first place. On the other hand, Gödel's theorem is not an a priori truth of nature! It is only as true as the set of axioms on which it is based. I, for one, do not accept the axiomatic basis of this theorem as relating to nature, since I believe(with Spinoza) that we are truly one with all of nature, not separable parts, but rather as manifestations of the holistic system that is the universe.

Thus, I believe that there is indeed 'complete knowledge'-a total underlying order. But I do not believe that we can ever reach this total knowledge, because it is infinite in extent. Still, I feel that it is our obligation, as scientists and philosophers, to pursue this objective knowledge of the universe, bit by bit, continually criticizing and rejecting what is scientifically invalid, and holding on to what is scientifically true.

According to Spinoza's view, in which there are no absolute lines of demarcation in the world between 'observer' and 'observed', and the leibnizian monad concept of the universe-an existent without actually separable parts-the tacit assumption that separates the human being from the rest of the universe must be rejected.

That is, rather than the atomistic view, in which the universe is said to be composed of many coupled, though separable parts-some of which are the collection of independent consciousnesses with their own free wills-the proponents of the unified field theory must view the world, with Spinoza, as a fully deterministic existent, that may exhibit an infinite manifold of intrinsic manifestations; yet where free will(actual individuality and separability) is only an apparent(illusory) feature, that is no more than a particular approximation for the oneness of the universe, as is the ripple of the pond example, discussed above.

Such a view of the universe, as a truly closed system, not only serves the purpose of providing an important heuristic function in the metaphysical approach toward a general theory of nature. It also has mathematical and logical consequences that do not follow from the type of universe that is taken to be an open set of things-a sum of individual parts.

Thus, the differences of these two metaphysical approaches-one in terms of a closed system and the other in terms of an open system-imply mutually exclusive features in the physical predictions that, in principle, could be tested further to verify one of these approaches or the other.

The philosophy of the unified field theory approach that we have been discussing bears a resemblance to the philosophies of many of the previous generations of scholars, even much earlier in the history of ideas than Spinoza and Leibniz. For example, some of Spinoza's views and his method of logical demonstration, can be traced to Moses Maimonides, who wrote his main philosophical treatise, *The Guide of the Perplexed,* in the 13th century.

I believe that these scholars saw the oneness of the universe to include humankind. Should society ever be able to accept this view, it could lead to a fully rational approach to science, as well as a higher ethical behaviour of all human beings-for it is a philosophic approach that implies humanism and a oneness with all of nature.

THE CONTINUOUS FIELD CONCEPT IN RELATIVITY

The debate between *atomism* and *continuity,* as fundamental ingredients in our comprehension of the universe, has been going on since the earliest periods in the Western and Oriental civilizations.

In ancient times, one asked the question, should one divide up any quantity of matter into finer and finer parts, would there be a final stopping point where one would reach the 'elementary atom' of matter? Or, would it be possible for this matter to be divided up *ad infinitum* in a continuous manner?

The atomists, who claim that there must be a stopping point, have been more successful in convincing most people of their view. Perhaps this is because the idea of discrete atoms more closely matches the human being's responses by way of the mind's perceptions of the world.

We tend to think of our surroundings as a collection of individual things. The continuous field concept is more abstract-it does not directly match our perceptual responses.

Perhaps, though, human beings should not expect scientific truth to reveal itself so simply, so as to yield its fundamental nature directly to our senses, or to our constructed instruments. I suspect that we are made in such a way that we require more subtle processes of discovery-using our percepts only for the purpose of extracting hints; then, in a next stage of comprehension, using our ability to reason in order to deduce the essence of the objective truths in science that are sought. Personally, I believe that this is the correct path toward a more complete comprehension of the universe.

With this approach, it is important to recognize that it is impossible for us to conclude any 'final truth' on the basis of our scientific investigations. I believe this to be the case because the amount of comprehension needed to complete any understanding of physical phenomena is infinite, while we are only finite human beings! In this regard it is interesting to recall Galileo's comment,

"There is not a single effect in Nature, not even the least that exists, that the most ingenious theorists can ever arrive at a complete understanding of it. We can only hope to approach the truth by successively matching more and more of the predictions of abstract theories that pertain to the observed facts of nature.

I think that this is one lesson that Faraday.taught when he introduced the field concept to oppose Newton's 'action-at-a-distance' concept, in a fundamental explanation of the material universe. For Faraday's initial findings about the unification of electricity and magnetism into the electromagnetic field of force was, according to his logic, only an initial step toward a fully unified field that, in principle, should exhibit all of the manifestations of matter in terms of continuity rather than atomism."

An important result discovered by Einstein during the early stages of relativity theory was that there is a fundamental incompatibility in maintaining the atomistic view within a theory that fully exploits the implications of the principle of relativity-which is the axiomatic basis of the theory of relativity.

Recall that this principle calls for the invariance of the laws of nature with respect to expressions of these laws in different frames of reference that are distinguishable only by virtue of relative motions. In this regard, there is a tacit assumption that motion refers to continuous change. In classical physics, this is a change of the three spatial coordinates with respect to the time coordinate.

In relativity, motion is defined more generally, as the continuous change of any of the four space-time coordinates of one frame of reference with respect to any of the four space-time coordinates of any other frame of reference.

This generalization follows because time is no longer an absolute coordinate, and space becomes space-time, as we have discussed previously. Such a generalization contains the classical expressions for motion, since, in a particular frame of reference in which the space-time interval appears as a pure time interval, the rate of change of space-time coordinates of a moving object, with respect to the space-time coordinates of that frame, appears as an ordinary velocity in that frame. In both classical and relativistic mechanics, however, motion, is a continuous entity.

The conservation laws give solutions that we identify with conserved quantities, such as energy, momentum and angular momentum. [We actually observe differences in these quantities rather than the quantities themselves.] The conservation laws follow, in this theory, from the invariance of the forms of all of the laws of nature with respect to continuous changes of the spatial and temporal coordinates.

The continuous changes are the translations along the space and time directions and the rotations in the(space-space) and(space-time) planes. Thus, the law of conservation of energy follows from the invariance(objectivity) of the laws of nature with respect to arbitrary, continuous shifts of the origin of the time coordinate axis. Similarly, the conservation of the three components of momentum(i.e. in each of the three spatial directions) follows from the objectivity of the laws of nature with respect to arbitrary, continuous shifts of the origin of the spatial coordinate system.

The law of conservation of angular momentum follows from the objectivity of the laws of nature with respect to the rotations in space. The translations and rotations are all of the continuous changes in space and time. [The proof that the laws of conservation relate to the invariance of the laws of nature with respect to continuous changes of the space and time coordinate systems follows from Noether's theorem, in the branch of mathematics called 'variational calculus'.]

If a real particle of matter is 'here'-therefore if it is not anywhere else-then how its intrinsic energy can content, its rest energy *Mc*, be defined in a continuous fashion throughout all of space and time? The modern atomic theory(quantum mechanics) avoids this difficulty by postulating that energy can only be emitted or absorbed in discrete quantities-'quanta'. This is as though there would be a filter placed between bits of interacting matter that only allows discrete quantitites of energy to be transferred between them.

That is, according to quantum theory, there can be transfers of energy, *E*1, *E*2, *E*3,..., but no energy between any two consecutive values, say between *E*2 and *E*3, can be transferred. No such quantum postulate is assumed in relativity theory, primarily because continuity is imposed there, in the definition of the continuous changes that distinguish different reference frames. This is in accordance with the symmetry imposed in space-time language by the principle of relativity.

It then follows that one cannot accept the discrete particle model of matter in a fully exploited theory of relativity. One must rather deal here with continuous fields-just as Faraday anticipated a century before Einstein. In a letter that Einstein wrote to David Bohm, in 1952, he said:

When one is not starting from the correct elementary concepts, if, for example, it is not correct that reality can be described as a continuous field, then all my efforts are futile, even though the constructed laws are the greatest simplicity thinkable.

Within the field theory the continuous variables relate to densities that are defined as continuous entities throughout space and time. To compare the predictions of the field theory with observed properties of matter, one must then sum these continuously varying fields over all of space(called 'integration', in calculus) so as to yield the numbers that are to be compared with the meter readings or any other means of recording measurements in physics experimentation.

We see here an example of the abstract approach, to explain the human being's observations of the world. As we have discussed previously, the empirical view taken by present day atomists tends to associate the data from experimentation with the theory itself-it identifies the descriptive level in science with the explanatory level, as is the view of naive realism.

On the other hand, the abstract approach postulates the existence of a theory that leads, by logical deductions only, to predictions that can then be compared with the data of experimentation. This is the case, in formal logic, of a universal leading to particulars. Should any of these particulars disagree with the facts of nature, the universal-the starting theoretical hypotheses for an explanation-would have to be altered in some way, or discarded altogether, and a new theory sought.

With the continuum view of field theory, the apparent discreteness of matter in the microscopic domain is, in reality, a high field concentration in a particular locality, while the field concentrations in other localities may be so weak that they appear to have zero amplitude. This would be analogous to describing particles like the ripples on the surface of a pond. With the atomistic view, any particular atom is separable from the rest of the system of atoms, while maintaining its individuality. However, in the pond example one cannot remove a ripple and study it on its own, as an individual thing-say, by viewing it under a microscope!

Still, one can indeed study the motion of an individual ripple in the pond-measure its energy and momentum, locate the maximum of its amplitude as a function of time, and so on.

But this 'ripple' is still an entity of the pond; it is a mode of behaviour of the entire pond that is, in principle, without parts. With the atomistic model, the whole is a sum of parts. With the 'holistic' model, there are no separate parts; rather, there is only a single continuous field that represents a closed physical system. This point of view to a physical model of the universe leads, in a natural way, to the Mach principle-a view of fundamental importance to the continued development of the theory of relativity.

Chapter 7

The Mach Principle

Ernst Mach was a philosopher-scientist whose writings, in the later years of the 19th century and the early years of the 20th century, had a profound influence on Einstein's development of the theory of relativity. Certain aspects of Mach's philosophy of science were at first accepted, but later rejected by Einstein.

Other aspects of Mach's scholarship had a strong influence throughout the entire development of relativity theory. In addition, and to my mind, of equal importance, was Mach's critical and sharp analysis of existing theories and ideas in physics and the philosophy of science, having a very beneficial influence on Einstein's scientific methodology.

On the method of science, Mach made the following comment. The history of science teaches that the subjective, scientific philosophies of individuals are constantly being corrected and obscured, and in the philosophy of constructive image of the universe which humanity gradually adopts, only the very strongest features of the thoughts of the greatest men are, after some lapse of time, recognizable. It is merely incumbent on the individual to outline as distinctly as possible the main features of his own view of the world.

In his comment that 'the very strongest features of the thoughts of men are after some lapse of time recognizable', I would contend that Mach is in agreement with the main thrust of Russell's philosophy of science-implying the existence of a real world, and the possibility of gradually approaching what it is that is true about this world, by a method of successive approximations of theoretical development of our comprehension. Mach's comment also reveals his openness about rejecting ideas about the world, if they can be found to be technically lacking or to have been scientifically refuted.

The latter is not the purely subjectivist view, that many attribute to Mach. From this quotation, we also see Mach's anti-dogmatic approach to scientific methodology. It is thus ironic that Mach's views have been rejected, totally, as dogmatic, by so many well-known scientists and philosophers. The reason for this is, in part, a rejection of Mach's positivistic philosophy of science. Similar to Berkeley's idealism, Mach supported the view that the only reality that one may talk about in a meaningful way must directly relate to the human being's sense perceptions. Einstein was indeed influenced by this approach in his initial studies of special relativity theory.

However, as we have discussed previously, he soon learned that this philosophical approach is not at all compatible with the stand of realism, that was to emerge from fully exploiting the principle of relativity. Rather than his positivistic philosophy, Mach's early influences on Einstein had more to do with his intellectually refreshing criticisms-mainly in his treatise, *The Science of Mechanics*.

Here, Mach presented a clear treatment of Newton's physical approach and a criticism of some of Newton's axioms of classical physics. Einstein agreed with his anti-dogmatic attitude-the idea that nothing in science should ever be accepted as an a priori truth. As it has been emphasized in more recent years by K.R. Popper, no 'truth' can be established once and for all, as an absolute element of objective knowledge. All that can be proven, scientifically, is a refutation of a scientific assertion, by conclusively demonstrating a logical and/or experimental incompatibility between implications of theoretical bases for natural phenomena and the full set of observational facts.

In *The Science of Mechanics,* Mach criticized Newton's attempts to establish the idea of an absolute space and time. He showed that equally one could derive the classical equations of motion of things from the relativistic view of space. Mach also offered a very interesting criticism of Newton regarding the manifestation of moving matter, called 'inertia', i.e. the resistance with which matter opposes a change in its state of motion(of constant speed or rest, relative to the reference frame of the source of the force that causes such change).

That is, it is more difficult for a tug boat to pull an ocean liner from the dock than to pull a sail boat, because the ocean liner has more 'inertial' mass than the sail boat.

In another example, if a large moving van, travelling at only 3 mph, should collide with a brick wall, it would probably disintegrate the wall. But if a man, walking at the same speed, should collide with the same wall, he would probably be thrown to the ground.

The difference in these situations is that the greater inertia of the van meant it had much more resistance to the action of the brick wall in stopping its motion than did the man, because of his much smaller inertial mass. The concept of inertia, as an intrinsic feature of moving matter, was first considered in ancient Greece, and referred to(by

Aristotle) as 'impetus'. But its correct quantitative features were not revealed until Galileo's discovery of his principle of inertia. This is the idea that a body at rest, or in constant rectilinear motion, relative to a stationary observer, would continue in this state of constant motion(or rest) forever, unless it should be compelled to change this state of motion by some external influence.

Newton quantified Galileo's principle of inertia with an explicit definition of this 'external influence' or 'force'. Newton asserted that if $F1$ is the magnitude of an external force, acting on a given body in causing it to accelerate at the rate, a_1, and if the force F_2 should cause the same body to accelerate at the rate a_2, then the ratio of the forces and ratio of 'caused' accelerations must be linearly related-whatever type of force that is used, i.e.

$$\frac{F_2}{F_1} = \frac{a_2}{a_1}$$

(This is an expression of Newton's second law of motion.)

Note also that, at this stage, there is no a priori reason, based on the statements of Galileo's principle of inertia, that this should be a linear relation. That is, Galileo's principle of inertia would not prohibit the possibility that this would be, e.g., a cubic relation,

$$F_2 / F_1 = (a_2 a_1)^3 .$$

But the empirical facts, regarding all of the observed forces in the classical period, including the electromagnetic forces, supported Newton's contention of linearity, as expressed in his second law of motion.

An equivalent way to express the linear relation between force and acceleration is in the usual form of Newton's second law of motion.

$$F = ma$$

The intrinsic property of matter, called 'inertial mass' and symbolized by the letter m, was then taken to be the constant of proportionality between the external force(the cause) and the resulting non-uniform motion(the effect) of the body acted upon, that is, its produced acceleration. The constant, m, cancels out when this law is expressed in the form of the ratio of two forces that act on the same body.

Mach noticed that the same empirical relation, $F = ma$, could be derived from a different conceptual interpretation of the inertial mass of matter. He argued as follows.

Suppose that two different bodies, with masses m_1 and m_2, are accelerated at the same rate by forces with magnitudes, F_1 and F_2. [An important example of this case is the acceleration of different bodies in the Earth's gravitational field.] In this case, the acceleration cancels in the ratio, and the ratio of forces is:

$$\frac{F_2}{F_1} = \frac{m_2}{m_1}$$

As in Newton's analysis, this relation may be re-expressed in the linear form

$$m = kF$$

where $k (= m_2 / F_2, \text{say})$ may be used as a standard for comparison with measures of the inertial masses of all other matter.

Equation may then be interpreted to say that the inertial mass, *m*, of a bit of matter is linearly proportional to the external force, *F*, that acts on it. Recall that, according to Newton's own definition, *F* is the total external force that acts on the matter(with mass *m*), whose source is all other matter of a closed system of matter.

Since the gravitational force, for example, has an infinite range, the actual physical system must, in principle, be taken as the entire universe. With this view, taken from a new interpretation of the empirically verified law of motion, *F* = *ma*, the inertial mass of any quantity of matter is *not* one of its intrinsic features. Rather, it is a measure of the coupling between this bit of matter and all other matter, of which it is an inseparable component.

This interpretation of the inertial mass, which was named by Einstein, 'the Mach principle', may be thought to be analogous to the resistance to the motion of a pellet of copper in a viscous medium, such as oil. This resistance is clearly dependent on the interaction between the metal pellet and the oil. If the oil(and all of the atmosphere) through which the pellet travels, would not be there, and if it would not be gravitationally affected by Earth or any other planet or star, then there would not be any resistance to the motion of this projectile.

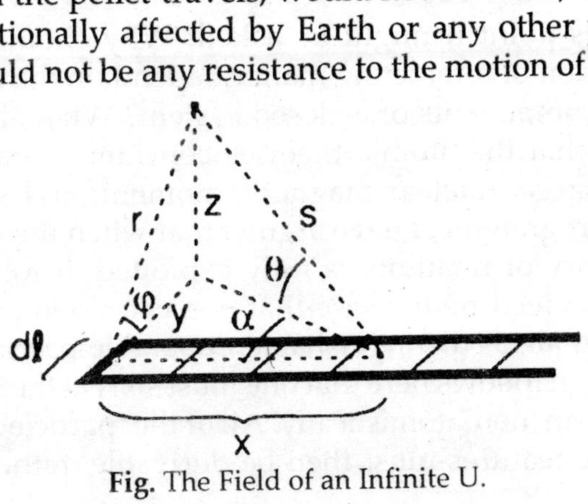

Fig. The Field of an Infinite U.

One might argue that is nothing more than a manipulation of formulae. The actual observables are represented for equal masses accelerated at different rates, or by for different masses accelerated at the same rate.

It is the ratio formula that relates directly to the data. From such empirical agreements with the ratio formulae, one could then speculate(with Newton) that inertial mass is a measure of a particular intrinsic property of a bit of matter, introduced theoretically as a constant of proportionality between a cause(the applied force, *F*) and the effect(the caused acceleration *a*) of the moving matter.

But one can equally speculate(with Mach), from the same empirical data, that the conceptual view of 'inertial mass', as a measure of dynamical coupling within a closed material system, is equally valid. With the latter view, inertia is not an intrinsic property of a bit of matter; it is rather a feature of an entire closed system of interacting material components.

Since all of the manifestations of matter are identified with the effects of forces, and the range of forces(at least the gravitational and the electromagnetic forces) is infinite in extent, the implications of Mach's argument are that any physical system must be considered to be closed. The total mass of the closed system is then not a simple sum of the masses of individual parts-since, with this view, each component mass is a function of the total force that acts where this mass is located. Thus, the dynamical states of motion of all other masses that make up the closed system affects the motion of any component mass of the physical system.

The interpretation of mass, according to Mach-the Mach principle-had a substantial influence on Einstein's development of the theory of relativity, especially its extension from 'special' relativity to 'general' relativity.

At this point it might be asked: Why should only the inertial manifestation of interacting matter depend on the mutual coupling of all of the components of a closed system? What about other features of matter that the atomic theories consider to be intrinsic, such as electric charge, nuclear magnetic moment, and so on? In my own research programme, I have argued that when the philosophical basis of the theory of relativity is fully exploited, it follows logically that this must indeed be the case-that one can no longer talk about 'free particles' at all, as distinguishable, separable parts.

Rather, it follows here that one must start at the outset with a single closed system, non-atomistically. All of the 'particle' features of matter-its intrinsic features-must then be derivable, rather than postulated, from the general theory, when certain approximations for the mathematical expression of the general theory can be made.

I have called this view 'the generalized Mach principle'. The mathematical structure of a theory of matter that incorporates the generalized Mach principle, is different from that of the present day theories of elementary particles. In contrast with the linear atomistic approaches of the current theories of matter, such as classical mechanics or the quantum theory, this is a continuum field theory, necessarily in terms of a non-linear structure.

In the former scheme, any number of solutions of the original laws forms a new, other possible solution. This is called the 'principle of linear superposition'. However, for a closed system, where the mathematical structure is non-linear, there can be no linear superposition.

It is interesting to note that two different aspects of Mach's philosophy have had opposite influences in 20th century physics. The philosophical basis of the present-day interpretation of quantum mechanics-to explain atomic phenomena and elementary particle physics, in accordance with the view of Niels Bohr-is quite close to Mach's positivistic approach. According to the latter view, all that can be said about the atoms of matter must necessarily be limited to the particular reactions of a large-sized(macroscopic) measuring apparatus to physical properties of micromatter.

It is basic in this view that, in principle, there is no underlying dynamical coupling that relates the observer(the macroapparatus) to the observed(the micromatter that is measured) in any exact sense. The observer(macroapparatus) is then said to respond to the different states of the observed(the micromatter) with various degrees of probability. The laws of nature, in this view, become laws of probability-a 'probability calculus'. This theory of micromatter is then fundamentally non-deterministic and it is based on the particularistic notion of atoms with imprecise trajectories, that is, without predetermined specifications of all of the parameters that characterize these trajectories simultaneously, with arbitrarily precise values.

This positivistic, non-deterministic view is in contrast with a different aspect of Mach's philosophy-the Mach principle-which implies that any material system must be closed at the outset.

The theory of relativity also implies this view. As with the quantum theory, one must start out with an 'observer' and the 'observed' to make any meaningful statement about a physical system. But in contrast with the quantum approach, this is a closed system, holistically described, without actual separable parts. 'Observer' is not a defined concept without the existence of the 'observed'-just as the concept 'early' is meaningless without the concept 'late'.

Further, and of equally important distinction, there is no intention in the theory of relativity to define 'observer' in anthropomorphic terms, or in terms of large scale quantitites with respect to small scaled 'observed' matter.

In this theory, observer-observed is taken to be a closed system. The hyphen is only used for purposes of convenience in particular situations where one can approximate the closed system by one component that is very weakly coupled to the rest of the system-sufficiently so that, under these circumstances, one can treat one of the coupled components to be an almost 'free' quantity of matter, and the other, the 'apparatus' that responds to this body.

But it is important with this view that, both from the philosophical and the mathematical standpoints, there is, in reality, no 'free matter', and no separated apparatus that looks down on it in a purely objective manner. The only objective entity, in this view, is a single, closed system, not composed of separate parts. This is the universe.

Question How do you feel the implications of a closed material system, according to the Mach principle, may relate to the human society?

Reply With the philosophy of science that takes any realistic system to be closed at the outset, there seems to me to be an overlap with particular views in sociology and psychology. Many physicists do not accept this holistic approach-indeed, the majority of contemporary physicists are atomists who adhere to the notion that any physical system is an open set of things, even though, according to the present-day view of quantum mechanics, the individual things are not said to have predetermined physical properties.

Similarly, I'm not sure that many of the contemporary sociologists and psychologists accept the view of a closed system underlying the human society.

Recall that 'closed system' does not refer to a sum of parts, such as a collection of human beings who are free of each other, except for their interactions.

It rather entails the society as a single entity that is without actual parts-even though it manifests itself as a system of weakly coupled parts, under the proper conditions.

It is nevertheless important in the descriptions of both the material, inanimate physical system, studied by the physical scientist, and the human society, studied by the social scientist, that the closed system does not predict observable features of the whole that are strictly a consequence of the intrinsic properties of 'parts'-it is rather a single entity without any actual parts.

TRANSITION TO GENERAL RELATIVITY

The combination of

- *The principle of relativity*-saying that the laws of nature must be independent of any frame of reference in which they may be expressed, by any particular observer, and the implication that they must be in terms of continuous field variables, and
- *Mach's principle*-implying that any real physical system in nature must be closed leads to the theory of general relativity.

The starting point of this generalization was Einstein's question: Why should the principle of relativity be confined only to comparisons of the laws of nature in inertial frames of reference, that is, frames that are in constant rectilinear relative motion? Should one not expect that the laws of nature would be the same in frames of reference that are in arbitrary types of relative motion?

His answer was that this must indeed be the case-since *all* motion is relative-as Galileo taught us in the 17th century! That is to say, if there should be a law of nature implying that Tamar accelerates relative to Jose, because there is some physical cause of her affected motion, then there must be an identical law, leading to the conclusion that Jose is accelerating relative to Tamar, as predicted by a corresponding cause-effect relation, but from Tamar's frame of reference rather than that of Jose. This is meant in the same way as we discussed previously.

If Tamar should deduce laws of nature from her observations of physical phenomena, within her own space-time reference frame, and if she should deduce the form of the laws of nature in Jose's reference frame, in terms of the space and time coordinates of that frame relative to her own, then she should find that the two expressions of a law of nature for any physical phenomenon in nature are in one-to-one correspondence, if these are indeed bona fide, objective laws of nature. We note that non-uniform motion is the only type of motion that can actually be experienced by matter, when it interacts with other matter.

This is because when one bit of matter reacts to other matter, this is due to a force that acts on it(by definition), causing a transfer of energy and momentum to it from the other matter. Force, *per se,* is the cause of non-uniform motion.

Thus, the case of uniform motion, which underlies special relativity theory, is an extrapolation from interaction to the ideal case, where the interaction turns off! That is to say, 'special' relativity is a limiting case that is, in principle, unreachable from the actual case of general relativity-where the only kind of relative motion between reference frames in which one compares the laws of nature is non-uniform.

Even so, one knows from the empirical evidence that the formulae of special relativity theory work perfectly well in describing a host of different sorts of experimental situations. To name a few of the important ones, it explained the results of

- The Michelson-Morley experiment,
- The Doppler effect for electromagnetic radiation,
- Changes in the momentum and energy of matter, as it moves at speeds close to the speed of light, that are different from the predictions of classical mechanics, and
- The well-known energy-mass relation, $E = Mc^2$ which has been well-verified in nuclear physics experimentation.

Still, there are properties of matter, observed in domains where special relativity provides an adequate description for other properties, that have not been explained within the formal expression of special relativity. An example is the set of physical properties associated with the inertia of matter, such as the magnitudes of the masses of the elementary particles-electron, proton, pi meson, and so on-and the empirical fact that they lie in a discrete spectrum of values.

Also not explained within special relativity is the feature of the gravitational force that it is only attractive, in the domain where Newton's theory of universal gravitation had been successful. This implies that the inertial mass of any bit of matter has only one polarity. This is in contrast with the electric charge of matter, that has two possible polarities, implying that electrical and magnetic forces can be either attractive or repulsive.

Let us now review some of the experimental data that have substantiated the formal expression of the theory of special relativity. The first is the *Michelson-Morley experiment*.

This was carried out in the last years of the 19th century. Its original aim was to measure the speed of the aether drift, relative to the Earth.(Everyone in that day believed aether to be the medium to conduct the flow of light.) The idea of this experiment was that as the Earth spins on its axis, it must be in motion relative to the stationary, all-pervasive aether.

As we observe the aether from a position on the spinning Earth, we should detect the effect of the aether on the speed of the propagating light waves, as they move in different directions relative to the Earth's axis. It was expected that the speed of light should be different in the directions parallel to the Earth's surface and perpendicular to it.

This is analogous to the motion of a small boat relative to a flowing river. If the river flows at V cm/sec in the eastwardly direction, then if the boat's speed in still water would be v cm/sec, Jill, who stands by the shore of the river would see that if the boat travelled eastwardly, its speed relative to her would be $v + V$ cm/sec; if it travelled westwardly in the river, its speed would be $v - V$ cm/sec relative to her, and if it crossed the stream in the northerly direction, its speed relative to the shore would be

$$(v^2 + V^2)^{\frac{1}{2}} \text{ cm/sec.}$$

In the Michelson-Morley experiment, v is the speed of the propagating light, if it should be in a vacuum, and V is the speed of the aether, relative to the Earth's surface, which is equivalent to the speed of the Earth relative to a stationary aether.

Consider a beam of light with a single frequency(called monochromatic), say the yellow light from a sodium lamp, as it is split along paths of equal lengths, but at different speeds. If the two beams should be brought back to the starting point with the use of mirrors, one should detect a loss of synchronization of their phases when they rejoin, if they were initially synchronized in phase. Of course, this is because it would take different times for the two light beams to traverse the same distances, if they travel at different speeds.

In the Michelson-Morley experiment, where one uses an 'interferometer' to study the comparison of the phases of the light beams, one having moved away from a source and back to it in a direction parallel to the surface of the earth, and the other moving away from the source and back to it in the direction perpendicular to the Earth's surface, one should then expect the recombined waves at the location of the source, because they are then out of phase, to yield an interference pattern. From this interference pattern one could then deduce the speed of the aether relative to the Earth's surface-with very high accuracy. Beside the magnitude of this speed, such an experiment would prove the existence of the aether to conduct light.

The result of this experiment was a null effect! That is, there was no interference pattern observed. One possible implication of this result was that there is no aether in the first place! It would be analogous to a vehicle travelling at a constant speed in a dry river bed-its speed seen to be the same in all directions-because there is no river there to alter it. If the observer could not see whether or not there is a flowing river, but he could observe that the vehicle travels at the same speed in the river bed in all directions, he would probably conclude that there is no river in the bed where it was supposed to be. In this way, the only reasonable conclusion from the Michelson-Morley experiment was that there was no aether there to conduct light. Such a resolution of the negative result from this experiment was inconceivable to most scientists of the late 19th and early 20th centuries-including Michelson himself. He firmly believed(with Maxwell, had he lived to see their experiment) that the radiation solutions of Maxwell's equations are an expression of the phenomenon of light as the vibrations of an aether-analogous to the interpretation of sound as the vibrations of a material medium.

Independently, Lorentz(in Holland) and Fitzgerald(in Ireland), interpreted the negative result of the Michelson-Morley experiment as a physical property of the aether, in that it interacts with physical instruments in such a way that they must contract by different amounts, depending on their direction of motion in the aether. Their formula for the amount of physical contraction, due to the force exerted by the aether on the measuring instruments(in the direction of its motion), was precisely the 'Lorentz transformation'.

Einstein's theory of special relativity, which was published several years after the Michelson-Morley experiment, though not referring directly to it, did explain their negative result in a logically consistent, theoretical fashion.

It was an explanation in terms of the idea that the aether is indeed a superfluous concept in physics-that it need not exist to explain the propagation of light. The explanation here was in terms of the relativity of space and time measures, that in turn led to the Lorentz transformations.

As we have discussed earlier, Einstein's explanation was along the lines of treating these transformations as scale changes of the space and time measures, when an observer expresses a law of nature in a frame of reference that is in motion relative to his own.

Thus they are no more than the 'translations' from the language appropriate to one reference frame to that of another, in order to preserve the form of the law(in accordance with the requirement of the principle of relativity.)

It is interesting to note that there is still controversy among historians of science on the question about whether or not Einstein was aware of the Michelson-Morley experimental result when he formulated special relativity theory.

It is also interesting that Michelson, himself, was not too happy about the explanation of his result with the theory of relativity-because of a feeling he had that this theory could not be correct! This points to advice that when analysing the history of science, one must take account of the fact that, far from being totally objective thinking machines, scientists are only human-along with all of the 'hang-ups' that go with this label!-such as irrational prejudices in science to fight off, as well as other emotional restraints, such as the near-omniscience that the scientist sometimes attributes to the leaders in his field!

The Doppler effect, in regard to the propagation of light, has a precedent in the classical description of wave motion. For example, when an ambulance or fire-engine passes, while sounding its siren, the sound heard has an increasing pitch(frequency), as it approaches, and a decreasing pitch as it departs; this phenomenon is called the Doppler effect, and was discovered in regard to the propagation of sound by Christian Doppler, in the 19th century. In a more quantitative investigation of this effect, it is found that the listener should move relative to the source of the sound *or,* if the source of the sound should move relative to the listener, in the opposite direction but with the same speed, the measured frequency shift would be slightly different in each of these cases.

The reason is that in the second case mentioned, where the source of sound is in motion relative to the listener, the oscillations of the density of the air molecules(which accounts for the phenomenon of sound) will 'bunch' the conducting, vibrating medium in the direction of motion of the source of the sound, while no such bunching effect occurs when it is the listener who is in motion relative to the source of the sound. Thus, the different Doppler effects for each of these cases is due to the existence of a medium whose role is to conduct the sound;(by compressing and rarefying, in time, to create the 'sound waves').

In the Doppler Effect it is the frequency of the measured sound wave that is changing. The quantitative effect that is predicted, when the listener moves away from the sound source, at v cm/sec, is:

$$f_m = f_s\left(1 - \frac{v}{c'}\right)$$

where f_m .is the frequency of the sound(measured in cycles per second by the listener), fs is the actual(proper) frequency of the sound emitted by the source and c_2 is the speed of sound in this particular medium.

In the second case, where it is the source of sound that is in motion relative to the listener,(in the opposite direction, but with the same magnitude of relative speed), the measured frequency is:

$$f_m = \frac{f_s}{\left(1 + \frac{v}{c'}\right)} \text{ cycles per second}$$

According to the binomial expansion, $1/(1+v/c')$ is practically the same as $(1-v/c')$, if the ratio v/c' is small compared with unity.(The actual difference depends on an infinite series of terms, in increasing powers of this ratio, starting with $(v/c')^2$.)

Thus, if the latter infinite series of terms can be neglected, these two Doppler effects would be numerically equal to each other. Nevertheless, these two effects are really measurably different, even if by a small amount when $v/c¢$ is small compared with unity(the usual case).

The Doppler effect in relativity theory, for radiation, is independent of whether the source of that radiation moves relative to the observer of it or vice versa. This is because the time coordinate is no longer an absolute measure-it is a subjective parameter, depending on the frame of reference in which it is expressed. Further, as we pointed out before, there is no medium(aether in this case) to conduct light-thus, there is no conducting substance to 'bunch' in this case, as happens with the propagation of sound.

The numerical predictions for the exact amount of frequency shift, according to the Doppler effect in relativity theory, has been experimentally confirmed.

Further, the theory predicts a Doppler effect when the frequency of light is measured in a direction that is perpendicular to the direction of the propagation of the light. This is called the 'transverse Doppler effect'. This effect has been experimentally confirmed in the case of electromagnetic radiation; it has no classical counterpart in the case of sound.

The changes in energy and momentum of matter, as a function of its increasing speed, according to the formulas of special relativity theory rather than classical physics, have been verified in numerous experimental tests.

For example, the designs of particle accelerators and the design of the mass spectrograph, are based on the dynamics of swiftly moving particles according to the predictions of the theory of special relativity.

The verification of the energy-mass relation, E=Mc, has been most outstanding in the observations of nuclear disintegration, such as nuclear fission, whereby heavy, unstable nuclei break up into a number of lighter ones, and nuclear fusion, such as the processes of interactions between light nuclei, giving rise to an emission of large amounts of energy, such as the processes of energy emission occurring in the Sun.

All of these data have been spectacularly supported by the predictions of the theory of special relativity. Still, according to what has been said earlier, this success can only be taken as an indication of the accuracy of the theory of special relativity as an approximation for a formal expression in general relativity.

For in spite of the remarkable accuracy of the formulae in special relativity in representing particular high energy data, it does not work in explaining other data in these same domains of measurement. For example, the inertial mass parameter must be inserted into the special relativity formulas, later to be adjusted to the experimental data. On the other hand, according to the full expression of the theory of general relativity, the mass parameter must be derivable from the general physical features of the closed system.

Question Is it possible that the theory of special relativity is no more than an accurate way of describing rapidly moving, electrically charged particles and electromagnetic radiation, while the theory of general relativity is an entirely different theory-a theory that has to do with gravitational forces and nothing more?

Reply There are critics of the general conceptual structure of relativity theory who argue that, because the formulae of the theory of special relativity 'work', there is no real justification for further generalization.

They say that the particular application of the mathematical apparatus of general relativity theory to describe successfully planetary motion and a few other phenomena having to do with gravitational forces, shows that this is an accurate description of gravitation, but it is not an explanation-indeed, that there is no need for further explanation!

This group of physicists takes the philosophic stand of logical positivism. That is, they look upon the formulae as no more than an economic way to describe the data, but they deny that they can come from any more basic ideas, such as an abstract law of nature that plays the role of a universal that underlies the empirical facts.

They claim that if some investigators have been lucky enough to derive predictions of the data from such claimed 'underlying universals' then it was purely coincidental(and perhaps a bit of chicanery!)

Of course, one can always take a set of formulae that happen to be 'working' at the time, in being able to fit particular data, and deny that there is anything more to talk about. But I would then claim that these people are at a dead end, as far as scientific progress is concerned!

With their point of view, they cannot make any progress in understanding the world-i.e. in deriving some of its features from first principles. All that they can do is to wait until more formulae appear on the scene, as will undoubtedly happen as science progresses. But these changes will not be discovered by the empiricists and positivists-unless a few miracles accidentally happen to drop the right formula in the right place at the right time! The changes in scientific understanding invariably come from investigations that are foundational in regard to the laws of nature. These, in turn, imply, by logical deduction, particulars such as the predictions of special types of data for corresponding experimental arrangements.

In the meanwhile, the empiricists and positivists will guard the existing formulae, as well as the meaning they attribute to them, with their lives! They will be the most fervent opponents to any real changes in the formulae-because, to this group of scientists, the data is the theory! Since the data cannot be wrong, they argue, their theory must be true, absolutely. In this philosophy, it is not usually recognized that whatever one says about the data must be expressed within the language of a particular theory, and this theory could be(in part, or totally) wrong! Indeed, we have seen this to be the case repeatedly throughout the history of science.

An example of the procedure in which one starts with a universal(a general law) and then derives particulars to be compared with the observations, is the theory of general relativity, with special relativity playing the role of a mathematical approximation that applies to some, but not all physical situations.

If an implied particular should be found to disagree with a single observed fact, then the entire theory would be challenged.

Chapter 8

Inertial Mass

Mechanics, as understood in post-Aristotelian physics, is generally regarded as consisting of kinematics and dynamics. Kinematics, a term coined by André-Marie Ampère, is the science that deals with the motions of bodies or particles without any regard to the causes of these motions.

Studying the positions of bodies as a function of time, kinematics can be conceived as a space-time geometry of motions, the fundamental notions of which are the concepts of length and time. By contrast, dynamics, a term probably used for the first time by Gottfried Wilhelm Leibniz, is the science that studies the motions of bodies as the result of causative interactions. As it is the task of dynamics to explain the motions described by kinematics, dynamics requires concepts additional to those used in kinematics, for "to explain" goes beyond "to describe."

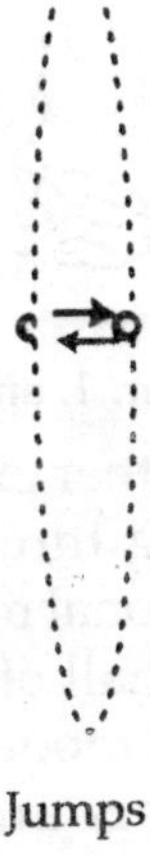

Fig. An Electron Jumps Through a Hoop.

The history of mechanics has shown that the transition from kinematics to dynamics requires only *one* additional concept—either the concept of mass or the concept of force. Following Isaac Newton, who began his *Principia* with a definition of mass, and whose second law of motion, in Euler's formulation $F = ma$, defines the force F as the product of the mass m and the acceleration a(acceleration being, of course, a kinematical concept), the concept of mass, or more exactly the concept of inertial mass, is usually chosen. The three fundamental notions of mechanics are therefore length, time, and mass, corresponding to the three physical dimensions L, T, and M with their units the meter, the second, and the kilogram. As in the last analysis all measurements in physics are kinematic in nature, to define the concept of mass and to understand the natures of mass are serious problems.

Fig. The Div-Meter, 1, and the Curl-Meter, 2 and 3.

These difficulties are further exacerbated by the fact that physicists generally distinguish among three types of masses, which they call inertial mass, active gravitational mass, and passive gravitational mass. For the sake of brevity we shall often denote them by m_i, m_a, and m_p, respectively. As a perusal of modern textbooks shows, contemporary definitions of these concepts are no less problematic than those published almost a century ago.

Today, as then, most authors define the inertial mass m_i of a particle as the ratio between the forced F acting on the particle and the acceleration a of the particle, produced by that force, or briefly as "the proportionality factor between a force and the acceleration produced by it." Some authors even add the condition that F has to be "mass-independent" (nongravitational), thereby committing the error of circularity.

The deficiency of this definition, based as it is on Newton's second law of motion $F = m; a$ is of cou.se its use of the notion of force. For if "force" is regarded as a primitive, that is, as an undefined term, then this definition defines an *ignotum per ignotius*; and if "force" is defined, as it generally is, as the product of acceleration and mass, then the definition is obviously circular.

The active gravitational mass m_a of a body, roughly defined, measures the strength of the gravitational field produced by the body, whereas its passive gravitational mass m_p measures the body's susceptibility or response to a given gravitational field. More precise definitions of the gravitational masses will be given later on.

Not all physicists differentiate between m_a and m_p. Hans C. Ohanian, for example, calls such a distinction "nonsense" because, as he says, "the equality between active and passive mass is required by the equality of action and reaction; an inequality would imply a violation of momentum conservation."

These comments are of course not intended to fault the authors of textbooks, for although it is easy to employ the concepts of mass it is difficult, as we shall see further on, to give them a logically and scientifically satisfactory definition. Even a genius such as Isaac Newton was not very successful in defining inertial mass! The generally accepted classification of masses into m_i, m_a, and m_p, the last two sometimes de..oted collectively by m_g for gravitational mass, gives rise to a problem.

Modern physics, as is well known, recognizes three fundamental forces of nature apart from gravitation—the electromagnetic, the weak, and the strong interactions.

Why then are noninertial masses associated only with the force of gravitation? True, at the end of the nineteenth century the concept of an "electromagnetic mass" played an important role in physical thought. But after the advent of the special theory of relativity it faded into oblivion. The problem of why only gravitational mass brings us to the forefront of current research in particle physics, for it is of course intimately related to the possibility, suggested by modern gauge theories, that the different forces are ultimately but different manifestations of one and the same force. From the historical point of view, the answer is simple.

Gravitation was the first of the forces to become the object of a full-fledged theory which, owing to the scalar character of its potential as compared with the vector or tensor character of the potential of the other forces, proved itself less complicated than the theories of the other forces. Although the notions of gravitational mass m_a and m_p differ conceptually from the notion of inertial mass m_i, their definitions, as we shall see later on, presuppose, implicitly at least, the concept of m_i. It is therefore logical to begin our discussion of the concepts of mass with an analysis of the notion of inertial mass.

There may be an objection here on the grounds that this is not the chronological order in which the various conceptions of mass emerged in the history of civilization and science. It is certainly true that the notion of "weight," i.e., $m_p g$, where g is the acceleration of free fall, and hence, by implication m_p, is much older than m_i. That weights were used in the early history of mankind is shown by the fact that the equal-arm balance can be traced back to the year 5000 b.c.

"Weights" are also mentioned in the Bible. In Deuteronomy, chapter 25, verse 13, we read: "You shall not have in your bag two kinds of weights, a large and a small ... a full and just weight you shall have." Or in Proverbs, chapter 11, verse 1, it is said: "A false balance is an abomination to the Lord, but a just weight is his delight."

But that "weight" is a force, given by $m_p g$, and thus involves the notion of gravitational mass could have been recognized only after Newton laid the foundations of classical dynamics, which he could not have done without introducing the concept of inertial mass.

Turning, then, to the concept of inertial mass we do not intend to recapitulate the long history of its gradual development from antiquity through Aegidius Romanus, John Buridan, Johannes Kepler, Christiaan Huygens, and Isaac Newton, which has been given elsewhere. Our intention here is to focus on only those aspects that have not yet been treated anywhere else.

One of these aspects is what has been supposed, though erroneously as we shall see, to be the earliest operational definition of inertial mass. But before beginning that discussion let us recall that, although Kepler and Huygens came close to anticipating the concept of m_i, it is Newton who has to be credited with having been the first to define the notion of inertial mass and to employ it systematically.

In particular, Galileo Galilei, as was noted elsewhere, never offered an explicit definition of mass. True, he used the term "massa," but only in a nontechnical sense of "stuff" or "matter." For him the fundamental quantities of mechanics were space, time, and momentum. He even proposed a method to compare the momenta("movimenti e lor velocità o impeti") of different bodies, but he never identified momentum as the product of mass and velocity. Richard S. Westfall, a prominent historian of seventeenth-century physics, wrote in this context: "Galileo does not, of course, clearly define mass.

His word *momento* serves both for our 'moment' and for our 'momentum,' and he frequently uses *impeto* for 'momentum.'" One of Galileo's standard devices to measure the *momenti* of equal bodies was to compare their impacts, that is, their *forze* of percussion."

It was therefore an anachronistic interpretation of Galileo's method of comparing momenta when the eminent mathematician Hermann Weyl wrote in 1927: "According to Galileo the *same* inert mass is attributed to two bodies if neither overruns the other when driven with equal velocities(they may be imagined to stick to each other upon colliding)."

This statement, which constitutes the first step of what we shall call "Weyl's definition of inertial mass," can be rephrased in more detail as follows: If, relative to an inertial reference frame *S*, two particles *A* and *B* of oppositely directed but equal velocities u_A and $u_B = -u_A$ collide inelastically and coalesce into a compound particle *A+B*, whose velocity u_{A+B} is zero, then the masses m^A and m_B, respectively, of these particles are equal.

In fact, if m_{A+B} denotes the mass of the compound particle, application of the conservation principles of mass and momentum, as used in classical physics, i.e., $m_A u_A + m_B m_B = M_{A+B} u_{A+B} = (m_A m_B) u_{A+B}$ shows that $u_B = -u_A$ and $u_{A+B} = 0$ imply $m_A = m_B$.

This test is an example of what is often called a "classificational measurement": Provided that it has been experimentally confirmed that the result of the test does not depend on the magnitude of the velocities u_A and u_B and that for any three particles *A*, *B*, and *C*, if $m_A = m_B$ and $m_B = m_C$ then the experiment also yields $m_A = m_C$ (i.e., the "equality" is an equivalence relation), it is possible to classify all particles into equivalence classes such that all members of such a class are equal in mass.

For a "comparative measurement," which establishes an order among these classes or their members, Weyl's criterion says: "That body has the larger mass which, at equal speeds, overruns the other." In other words, m_A is larger than m_B, or $m_A > m_B$, if $u_A = -u_B$ but $u_{A+B} \neq 0$ and sign u_A = sign u_{A+B}. To ensure that the relation "larger" thus defined is an order relation it has to be experimentally confirmed that it is an asymmetric and transitive relation, i.e., if $m_A > m_B$ then $m_B > m_A$ does not hold, and if $m_A > m_B$ and $m_B > m_C$ have been obtained then $m_A > m_C$ will also be obtained for any three particles *A*, *B*, and *C*. Since for $u_A = -u_B$ equation(1.2) can be written $m_A - m_B = (u_{A+B} / u_A) m_{A+B}$ the condition sign u_A = sign u_{A+B} shows that the coefficient of m_{A+B} is a positive number and, hence, $m_A > m_B$, it being assumed, of course, that all mass values are positive numbers. The experimentally defined relation ">" therefore coincides with the algebraic relation denoted by the same symbol.

Finally, to obtain a "metrical measurement" the shortest method is to impose only the condition $u_{A+B} = 0$ so that equation reduces to $m_A / m_B = -u_B / u_A$. Hence, purely kinematic measurements of u_A and u_B determine the mass-ratio m_A/m_B.

Choosing, say, m_B as the standard unit of mass($m_B = 1$) determines the mass m_A of any particle A unambiguously. Weyl called this quantitative determination of mass "a definition by abstraction" and referred to it as "a typical example of the formation of physical concepts." For such a definition, he pointed out, conforms to the characteristic trait of modern science, in contrast to Aristotelian science, to reduce qualitative determinations to quantitative ones, and he quoted Galileo's dictum that the task of physics is "to measure what is measurable and to try to render measurable what is not so yet."

Weyl's definition of mass raises a number of questions, among them the philosophical question of whether it is really a definition of inertial mass and not only a prescription of how to measure the magnitude of this mass.

It may also be asked whether it does not involve a circularity; for the assumption that the reference frame S is an inertial frame is a necessary condition for its applicability, but for the definition of an inertial system the notion of force and, therefore, by implication, that of mass may well be indispensable.

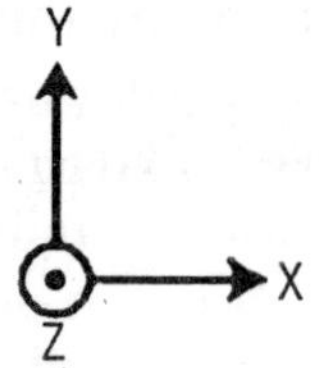

Fig. The Coordinate System used in the following Examples.

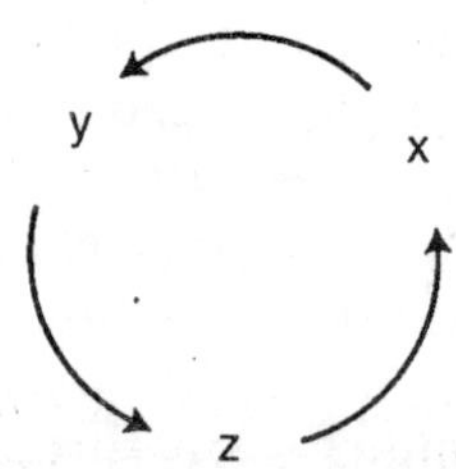

Fig. A Cyclic Permutation of *x*, *y*, and *z*.

Not surprisingly, Weyl's definition seems never to have been criticized in the literature on this subject, for the same questions have been discussed in connection with the much better-known definition of mass that Ernst Mach proposed about sixty years earlier. In fact, these two definitions have much in common. The difference is essentially only that Weyl's definition is based, as we have seen, on the principle of the conservation of momentum while Mach's rests on the principle of the equality between action and reaction or Newton's third law. But, as is well known, both principles have the same physical content because the former is only a time-integrated form of the latter.

Although Mach's definition of inertial mass is widely known, we shall review it briefly for the convenience of the reader. For Mach, just as for Weyl six decades later, the task of physics is "the abstract quantitative expression of facts." Physics does not have to "explain" phenomena in terms of purposes or hidden causes, but has only to give a simple but comprehensive account of the relations of dependence among phenomena. Thus he vigorously opposed the use of metaphysical notions in physics and criticized, in particular, Newton's conceptions of space and time as presented in the *Principia*.

Concering Newton's definition of mass Mach declared: "With regard to the concept of 'mass,' it is to be observed that the formulation of Newton, which defines mass to be the quantity of matter of a body as measured by the product of its volume and density, is unfortunate. As we can only define density as the mass of a unit of volume, the circle is manifest." In order to avoid such circularity and any metaphysical obscurities Mach proposed to define mass with an operational definition. It applies the dynamical interaction between two bodies, called A and B, that induce in each other opposite accelerations in the direction of their line of junction.

If $a_{A/B}$ denotes the acceleration of A owing to B, and $a_{B/A}$ the acceleration of B owing to A, then, as Mach points out, the ratio $-a_{B/A}/a_{A/B}$ is a positive numerical constant independent of the positions or motions of the bodies and defines what he calls the mass ratio

$$m_{A/B} = -a_{B/A}/a_{A/B}.$$

By introducing a third body C, interacting with A and B, he shows that the mass-ratios satisfy the transitive relation

$$m_{A/B} = m_{A/C}\, m_{C/B}$$

and concludes that each mass-ratio is the ratio of two positive numbers, i.e.,

$$m_{A/B} = m_A/m_B,\quad m_{A/C} = m_A/m_C,$$

and

$$m_{C/B} = m_C/m_B.$$

Finally, if one of the bodies, say A, is chosen as the standard unit of mass($m_A = 1$), the masses of the other bodies are uniquely determined.

Mach's identification of the ratio of the masses of two interacting bodies as the negative inverse ratio of their mutually induced accelerations is essentially only an elimination of the notion of force by combining Newton's third law of the equality between action and reaction with his second law of motion. In fact, if F_{AB} is the force exerted on A by B and F_{BA} the force exerted on B by A, then according to the third law $F_{AB} = -F_{BA}$. But according to the second law

$$F_{AB} = m_A a_{A/B}$$

and

$$F_{BA} = m_B a_{B/A}.$$

Hence,

$$m_A a_{A/B} = -m_B a_{B/A}$$

or

$$m_{A/B} = m_A/m_B = -a_{B/A}/a_{A/B},$$

as stated by Mach, and the mass-ratio $m_{A/B}$ is the ratio between two inertial masses. Thus we see that Mach's operational definition is a definition of *inertial* masses.

We have briefly reviewed Mach's definition not only because it is still restated in one form or another in modern physics texts, but also, and more importantly, because it is still a subject on which philosophers of science disagree just as they did in the early years of the century. In fact, as we shall see, recent arguments *pro* or *contra* Mach's approach were first put forth a long time ago, though in different terms.

For example, in 1910 the philosopher Paul Volkmann declared that Mach's "phenomenological definition of mass", as he called it, contradicts Mach's own statement that the notion of mass, since it is a fundamental concept("Grundbegriff"), does not properly admit any definition because we deprive it of a great deal of its rich content if we confine its meaning solely to the principle of reaction.

On the other hand, the epistemologist and historian of philosophy Rudolf Thiele declared that "one can hardly overestimate the merit that is due to Mach for having derived the concept of mass without any recourse to metaphysics. His work is also important for the theory of knowledge, since it provides for the first time, an *immanent* determination of this notion without the necessity of transcending the realm of possible experience."

As noted above, many textbooks define inertial mass m_i as the ratio between the force F and the acceleration a in accordance with Newton's second law of motion, which in Euler's formulation reads F

= $m_i a$. Further, they often suppose that the notion of force is immediately known to us by our muscular sensation when overcoming the resistance in moving a heavy body.

But there are also quite a few texts on mechanics that follow Mach, even though they do not refer to him explicitly, and introduce m_i in terms of an operational definition based either on Newton's third law, expressing the equality of action and reaction, or on the principle of the conservation of linear momentum. It is therefore strange that the prominent physicist and philosopher of physics, Percy Williams Bridgman, a staunch proponent of operationalism and probably the first to use the term "operational definition," never even mentioned Mach's operational definition of mass in his influential book *The Logic of Modern Physics*, although his comments on Mach's cosmological ideas clearly show that he had read Mach's writings.

Instead, like many physicists and philosophers of the late nineteenth century, among them James Clerk Maxwell and Alois Höfler, Bridgman introduced "mass" essentially in accordance with Newton's second law, but put, as he phrased it, "the crude concept [of force] on a quantitative basis by substituting a spring balance for our muscles, or instead of the spring balance … any elastic body, and [we] measure the force exerted by it in terms of its deformation." After commenting on the role of force in the case of static systems Bridgman continued:

We next extend the force concept to systems not in equilibrium, in which there are accelerations, and we must conceive that at first all our experiments are made in an isolated laboratory far out in empty space, where there is no gravitational field. We here encounter a new concept, that of mass, which as it is originally met is entangled with the force concept, but may later be disentangled by a process of successive approximations.

The details of the various steps in the process of approximation are very instructive as typical of all methods in physics, but need not be elaborated here. Suffice it to say that we are eventually able to give to each rigid material body a numerical tag characteristic of the body such that the product of this number and the acceleration it receives under the action of any given force applied to it by a spring balance is numerically equal to the force, the force being defined, except for a correction, in terms of the deformation of the balance, exactly as it was in the static case.

In particularly, the relation found between mass, force, and acceleration applies to the spring balance itself by which the force is applied, so that a correction has to be applied for a diminution of the force exerted by the balance arising from its own acceleration. We have purposely quoted almost all of what Bridgman had to say about the definition of mass in order to show that the definition of mass *via* an operational definition of force meets with not inconsiderable difficulties. Nor do his statements give us any hint as to why he completely ignored Mach's operational definition of mass.

In the late 1930s Mach's definition was challenged as having only a very limited range of applicability insofar as it fails to determine unique mass-values for dynamical systems composed of an arbitrary number of bodies. Indeed, C. G. Pendse claimed in 1937 that Mach's approach breaks down for any system composed of more than four bodies. Let us briefly outline Pendse's argument. If in a system of n bodies a_k denotes, in vector notation, the observable induced acceleration of the kth body and $u_{kj}(j \neq k)$ the observable unit vector in the direction from the kth to the jth body, then clearly

$$a_k = \sum_{j=1}^{n} a_{kj} u_{kj} \quad (k = 1.2, \ldots, n),$$

where $\alpha_{kj}(\alpha_{kk} = 0)$ are $n(n-1)$ unknown numerical coefficients in $3n$ algebraic equations. However, these coefficients, which are required for the determination of the mass-ratios, are uniquely determined only if their number does not exceed the number of the equations, i.e., $n(n-1) \leq 3n$, or $n \leq 4$.

Pendse also looked into the question of how this result is affected if the dynamical system is observed at r different instants. Again using simple algebra he arrived at the conclusion that "if there be more than *seven* particles in the system the observer will be unable to determine the ratios of the masses of the particles ..., however large the number of instants, the accelerations pertaining to which are considered, may be."

Pendse's conclusions were soon challenged by V. V. Narlikar on the grounds that the Newtonian inverse-square law of gravitation, if applied to a system of n interacting massive particles, makes it possible to assign a unique mass-value $m_k(k = 1, 2, \ldots, n)$ to each individual particle of the system. For according to this law, the acceleration a_K of

the kth particle satisfies the equation $a_k = \sum_{\substack{j=1 \\ j \neq 1}}^{n} G m_j \mathrm{r}_{jk} / |\mathrm{r}_{jk}|^3$,

where G is the constant of gravitation and r_{jk} is the vector pointing from the position of m_k to the position of m_j. Since all accelerations $a_k(k = 1, 2, \ldots, n)$ and all r_{jk} are observable, "all the masses become known in this manner."

It should be noted, however, that Narlikar established this result for active gravitational masses, for the m_j in the above equations are those kinds of masses, and not for inertial masses, which we have seen were the *definienda* in Pendse's approach. It is tempting to claim that this difficulty can be resolved within Mach's conceptual framework by an appeal to his *experimental proposition,* which says: "The mass-ratios of bodies are independent of the character of the physical states(of the bodies) that condition the mutual accelerations produced, be those states electrical, magnetic, or what not; and they remain, moreover, the same, whether they are mediately or immediately arrived at." Hence one may say that the interactions relative to which the mass-ratios are invariant also include gravitational interactions although these were not explicitly mentioned by Mach. However, this interpretation may be questioned because of Mach's separate *derivation* of the measurability of mass by weight. As this derivation illustrates, quite a few problematic issues appertaining to Mach's treatment of mass would have been avoided had he systematically distinguished between inertial and active or passive gravitational mass.

A serious difficulty with Mach's definition of mass is its dependence on the reference frame relative to which the mutually induced accelerations are to be measured. Let us briefly recall how the mass-ratio $m_{A/B}$ of two particles *A* and *B* depends on the reference frame *S*. In a reference frame *S*2, which is moving with an acceleration *a* relative to *S*, we have by definition

$$m'_{A/B} = -a'_{B/A}/a'_{A/B} = -(a_{B/A} - a)/(a_{A/B} - a)$$

so that

$$m'_{A/B} = m_{A/B}[1-(a/a_{B/A})]/[1-(a/a_{A/B})] \neq m_{A/B} (\text{for } a \neq 0).$$

Thus in order to obtain uniquely determined mass-values, Mach assumed, tacitly at least, that the reference frame to be used for the measurement of the induced accelerations is an inertial system However, such a system is defined by the condition that a "free" particle(i.e., a particle not acted upon by a force) moves relative to it in uniform rectilinear motion.

This condition involves, as we see, the notion of force, which Mach defined as "the product of the mass-value of a body times the acceleration induced in that body." Hence, Mach's definition involves a logical circle. Nevertheless, in the early decades of the twentieth century Mach's definition of mass, as an example of his opposition to the legitimacy of metaphysics in scientific thought, enjoyed considerable popularity, especially among the members of the Viennese Circle founded by Moritz Schlick.

Repudiating Kantian apriorism, logical positivists and scientific empiricists stressed the importance of the logical analysis of the fundamental concepts of physical science and often regarded Mach's definition of mass as a model for such a programme. A drastic change occurred only after the 1950s when the positivistic philosophy of science became a subject of critical attack.

One of the most eloquent critics was the philosopher Mario Bunge. According to Bunge, Mach committed a serious error when he "concluded that he has *defined* the mass concept in terms of observable(kinematic) properties," for, "Mach confused 'measuring' and 'computing' with 'defining.'" In particular, the equation

$$m_A/m_B = -a_{B/A}/a_{A/B'}$$

which establishes an equality between two expressions that differ in meaning—the left-hand side expressing "the inertia of body *A* relative to the inertia of body *B*" and the right-hand side standing for a purely kinematical quantity—cannot be interpreted, as Mach contended, as having the meaning of a definition.

It is a numerical, but not a logical, equality and "does not authorize us to eliminate one of the sides in favour of the other." In a similar vein Renate Wahsnerand Horst-Heino von Borzeszkowski rejected Mach's definition on the grounds that "the real nature"("das Wesen") of mass cannot be obtained by merely quantitative determinations. Moreover, they charged Mach, as Ludwig Boltzmann had done earlier, with contradicting his own precept that a mechanics that transcends experience fails to perform its proper task.

Mach's definition, based as it is on the interaction between two mutually attracting bodies, has not been proved to be universally valid for all bodies dealt with in mechanics and his claim that the "experimental propositions" do not go beyond experience is confuted by the fact that they presuppose all principles of mechanics.

Similarly, in a recent essay on operational definitions Andreas Kamlah rejects the claim that the concept of mass can in all cases be defined in a kinematical language containing only the notions of position, time, and velocity(or acceleration).

He also argues that "Mach's definition is not a definition in the proper sense ... [for] it yields the values of mass only for bodies which just by chance collide with other bodies. All other values of that function remain undetermined."

In contrast to the preceding unfavorable criticisms(and many others could have been recounted), Mach's definition was defended, at least against two major objections, by Arnold Koslow. The two objections referred to concern the restricted applicability of the definition and its noninvariance relative to different reference frames.

Koslow's main argument against the former objection contends that the third experimental proposition has not been taken into account. For according to this proposition the mass-ratios are independent of whether the mutual accelerations are induced by "electric, magnetic, or what not" interactions.

Hence, as Koslow shows in mathematical detail, by performing the definitional operations with respect to different kinds of interactions, the number of the equations can be sufficiently increased to ensure the uniqueness of the mass-ratios for any finite number of particles. Concerning the latter objection, Koslow justified Mach's contention that "the earth usually does well enough as a reference system, and for larger scaled motions, or increased accuracy, one can use the system of the fixed stars."

An operational definition of inertial mass, which unlike Mach's definition seems to be little known even among experts, is the so-called "tabletop definition" proposed in 1985 by P. A. Goodinson and B. L. Luffman.

Unlike Mach's and Weyl's definitions of m_i, which are based, as we have seen, on Newton's third law, the Goodinson-Luffman definition is based on Newton's second law, which, in Euler's formulation, says that force is the product of mass and acceleration. However, as the notion of force(or of weight or of friction) as used in this definition is made part of the operational procedure, an explicit definition is not required so that from the purely operational point of view they seem to have avoided a logical circularity.

Goodinson and Luffman call their definition of m_i a "table-top definition" because it involves the measurement of the acceleration a_B of a body B that is moving on a horizontal table—on "a real table, not the proverbial 'infinitely smooth table.'"

The motion of B is produced by means of a(weightless) string that is attached to B, passes over a(frictionless) pulley fixed at the end of the table, and carries a heavy weight W on its other end. At first the acceleration a_0 of a standard body B_0, connected via the string with an appropriate weight W_0, is measured. Measurements of distance and time are of course supposed to have been operationally defined antecedently, just as in the operational definitions by Mach or by Weyl.

The procedure of measuring the acceleration a is repeated for a body B and also for weights W that differ from W_0. A plot of a against a_0 shows that

$$a = ka_0 + c,$$

where k and c are constants. Repetition of the whole series of measurements with a different table again yields a linear relation

$$a = ka_0 + d$$

with the same slope k but with a constant d that differs from c.

This shows that the intercepts c and d are table-dependent whereas the slope k is independent of the roughness or friction caused by the table. A series of such measurements for bodies $B_q (q = 1, 2, \ldots)$ yields a series of straight-line plots, one plot for each a_q against a_0 with slope k_q. These slopes are seen to have the following properties: if B_q is "heavier than" B_p then

$$k_q < k_p$$

$$1/k_q + 1/k_p = 1/k_{p+q}$$

and where k_{q+p} is the slope obtained when B_q and B_p are combined. The inertial mass $m_i(B_q)$ of a body B_q, with respect to the standard body B_0, is now defined by

$$m_i(B_q) = 1/k_q.$$

In the sequel to their paper Goodinson and Luffman prove that equations and are independent of the choice of the standard body B_0, and that

$$m_i(B_1) = m_i(B_2)$$

and

$$m_i(B_2) = m_i(B_3)$$

imply

$$m_i(B_1) = m_i(B_3)$$

for any three bodies B_1, B_2, and B_3, independently of the choice of B_0. In addition to this transitivity of mass, the additivity of mass is obviously assured because of.

That in spite of the fundamental differences noted above the table-top definition converges to Mach's definition under certain conditions can be seen as follows. For two arbitrary bodies B_1 and B_2 with inertial masses

$$m_i(B_1) = k^{-}{}_1$$

and

$$m_i(B_2) = k^{-}{}_2,$$

the plots of their respective accelerations a_1 and a_2 with respect to B_0 are

$$a_1 = \left[m_i(B_1)\right]^{-1} a_0 + c_1$$

and

$$a_2 = \left[m_i(B_2)\right]^{-1} a_0 + c_2.$$

Hence

$$m_1(B_1)a_1 = m_i(B_2)a_2 + c_{12}.$$

where

$$c_{12} = m_i(B_1)c_1 - m_i(B_2)c_2.$$

Experience shows that the quantity $|c_{12}$ is table-dependent and approaches zero in the case of a perfectly smooth table. In the limit, which agrees with the Machian definition of the mass-ratio of two bodies as the inverse ratio of their accelerations(the minus sign being ignored)?

Yet in spite of this agreement the table-top definition is proof against the criticism leveled against Mach's definition as being dependent on the reference frame. In fact, if an observer at rest in a reference frame S graphs the plot for a body B_1 with respect to B_0 in the form

$$a_1 = [m_1(B_1)]^{-1} a_0 + c_1.$$

Then an observer at rest in a reference frame $S2$ that moves with an acceleration a relative to S(in the direction of the accelerations involved) will write

$$a'_1 = [m'_i(B_1)]^{-1} a'_0 + c'_1.$$

But since

$$a'_1 = a_1 - a$$

and

$$a'_0 = a' - a,$$

clearly

$$a'_1 = [m'_i(B_1)]^{-1} a_0 + c''_1.$$

where

$$c''_1 = c'_1 + a[1 - m'_i(B_1)]^{-1}$$

Hence, the plot of a_1 against a_0 has the slope $[m'_i(B_1)]''$, which shows, if compared with, that

$$m_i(B_1) = m'_i(B_1)$$

since m_i is defined only by the slope. Thus, both observers obtain the same result when measuring the inertial mass of the body B_1. Of course, this conclusion is valid only within the framework of classical mechanics and does not hold, for instance, in the theory of relativity.

The range of objects to which an operational definition of inertial mass, such as the Goodinson-Luffman definition, can be applied is obviously limited to medium-sized bodies. One objection against operationalism raised by philosophers of the School of Scientific Empiricists, an outgrowth of the Viennese School of Logical Positivists, is that quite generally no operational definition of a physical concept, and in particular of the concept of mass, can ever be applied to all the objects to which the concept is attributed. Motivated by the apparently unavoidable circularity in Mach's operational definition of mass they preferred to regard the notion of mass as what they called a partially interpreted theoretical concept.

A typical example is Rudolf Carnap's discussion of the notion of mass. The need to refer to different interactions or different physical theories when speaking, e.g., of the mass of an atom or of the mass of a star, led him to challenge the operational approach. Instead of saying that there are various concepts of mass, each defined by a different operational procedure, Carnap maintained that we have merely one concept of mass. "If we restrict its meaning [the meaning of the concept of mass] to a definition referring to a balance scale, we can apply the term to only a small intermediate range of values.

We cannot speak of the mass of the moon. ... We should have to distinguish between a number of different magnitudes, each with its own operational definition. ... It seems best to adopt the language form used by most physicists and regard length, mass and so on as theoretical concepts rather than observational concepts explicitly defined by certain procedures of measurement."

Carnap's proposal to regard "mass" as a theoretical concept refers of course to the dichotomization of scientific terms into observational and theoretical terms, an issue widely discussed in modern analytic philosophy of science. Since, generally speaking, physicists are not familiar with the issue, some brief comments, specially adapted to our subject, may not be out of place.

It has been claimed by philosophers of science that physics owes much of its progress to the use of theories that transcend the realm of purely empirical or observational data by incorporating into their conceptual structure so-called theoretical terms or theoretical concepts.(We ignore the exact distinction between the linguistic entity "term" and the extralinguistic notion "concept" and use these two words as synonyms.)

In contrast to "observational concepts," such as "red," "hot," or "iron rod," whose meanings are given ostensively, "theoretical concepts," such as "potential," "electron," or "isospin," is not explicitly definable by direct observation. Although the precise nature of a criterion for observability or for theoreticity has been a matter of some debate, it has been generally agreed that terms, obtaining their meaning only through the role they play in the theory as a whole, are theoretical terms. This applies, in particular, to terms, such as "mass," used in axiomatizations of classical mechanics, such as proposed by H. Hermes, H. A. Simon, J.C.C. McKinsey *et al.*, S. Rubin and P. Suppes, or more recently by C. W. Mackey, J. D. Sneed, and W. Stegm üller.

In these axiomatizations of mechanics "mass" is a theoretical concept because it derives its meaning from certain rules or postulates of correspondence that associate the purely formally axiomatized term with specific laboratory procedures. Furthermore, the purely formal axiomatization of the term "mass" is justified as a result of the confirmation that accrues to the axiomatized and thus interpreted theory as a whole and not to an individual theorem that employs the notion of mass.

It is for this reason that Frank Plumpton Ramsey seems to have been the first to conceive "mass" as a theoretical concept when he declared in the late 1920s that to say "'there is such a quality as mass' is nonsense unless it means merely to affirm the consequences of a mechanical theory." Ramsey was also the first to propose a method to eliminate theoretical terms of a theory by what is now called the "Ramsey sentence" of the theory.

Briefly expressed, it involves logically conjoining all the axioms of the theory and the correspondence postulates into a single sentence, replacing therein each theoretical term by a predicate variable and quantifying existentially over all the predicate variables thus introduced. This sentence, now containing only observational terms, is supposed to have the same logical consequences as the original theory. The term "mass" has been a favourite example in the literature on the "Ramsey sentence."

Carnap proposed regarding "mass" as a theoretical concept, as we noted above, because of the inapplicability of one and the same operational definition of mass for objects that differ greatly in bulk, such as a molecule and the moon, and since different definitions assign different meanings to their definienda, the universality of the concept of mass would be untenable.

However, this universality would also be violated if the mass, or rather masses, of one and the same object are being defined by operational definitions based on different physical principles. This was the case, for instance, when Koslow suggested employing different kinds of interactions in order to rebut Pendse's criticism of Mach's definition as failing to account for the masses of arbitrarily many particles.

Even if in accordance with Mach's "experimental proposition" the numerical values of the thus defined masses are equal, the respective concepts of mass may well be different, as is, in fact, the case with inertial and gravitational mass in classical mechanics, and one would have to distinguish between, say, "mechanical mass"(e.g., "harmonic oscillator mass"), "Coulomb law mass," "magnetic mass," and so on.

The possibility of such a differentiation of masses was discussed recently by Andreas Kamlah when he distinguished between "energyprinciple mass"("Energiesatz-Masse") and "momentum-principle mass" ("Impulssatz-Masse"), corresponding to whether the conservation principle of energy or of momentum is being used for the definition.

Thus, according to Kamlah, the energy-principle masses $m_k (k = 1, \ldots, n)$ of n free particles can be determined by the system of equations

$$\frac{1}{2}\sum_{k=1}^{n} m_k u_k^2 (t) = c,$$

where $u_k(t_j)$ denotes the velocity of the kth particle at the time $t_j (j = 1, \ldots, r)$ and c is a constant. In the simple case of an elastic collision between two particles of velocities u_1 and u_2 before, and u'_1 and u'_2 after, the collision, the equation

$$\frac{1}{2} m_1 u_1^2 + \frac{1}{2} m_2 u_2^2 = \frac{1}{2} m_1 u'^2_1 + \frac{1}{2} m_2 u'^2_2$$

determines the mass ratio

$$m_1 / m_2 = \left(u'^2_2 - u_2^2\right) / \left(u_1^2 - u'^2_1\right).$$

The momentum-principle masses μ_k of the same particles are determined by the equations

$$\sum_{k-1}^{n} \mu_k u_k \left(t_j\right) = P,$$

where P, the total momentum, is a constant. In the simple case of two particles, the equation

$$\mu_1 u_1 + \mu_2 u_2 = \mu_2 u_1' + \mu_2 u_2'$$

determines the mass-ratio,

$$\mu_1 / \mu_1 = \left(\mu'_2 - u_2\right) / \left(u_1 - u_1'\right).$$

The equality between m_1 / m_2 and μ_1 / μ_2 cannot be established without further assumptions, but as shown by Kamlah, it is sufficient to postulate the translational and rotational invariance of the laws of nature.

More specifically, this equality is established by use of the Hamiltonian principle of least action or, equivalently, the Lagrangian formalism of mechanics, both of which, incidentally, are known to have a wide range of applicability in physics. The variational principle $\delta \int L / dt = 0$ implies that the Lagrangian function

$$L = L(x_1, \ldots, x_n, u_1, \ldots, u_n, t)$$

satisfies the Euler-Lagrange equation

$$\sum_j \left(\frac{\partial^2 L}{\partial u_i \partial u_j} \dot{u}_j + \frac{\partial^2 L}{\partial u_i \partial x_j} u_j \right) - \frac{\partial L}{\partial x_i} = 0 \qquad \dot{u}_j = du_j / dt.$$

By defining generalized masses $m_{ij}(u_1, \ldots, u_n)$ by

$$m_{ij} = -\delta L / \delta u_i \, \delta u_j,$$

and masses m_i, assumed to be constant, by $m_{ij} = m_i \, \delta_{ij}$, and taking into consideration that the spatial invariance implies,

$$\sum_i \partial L / \partial x_i, = 0$$

Kamlah shows that the Euler-Lagrange equation reduces to

$$\sum_i \partial L / \partial u_i = P = \text{const.},$$

where $\partial L / \partial u_i = m_i u_i$. Comparison with equation yields

$$m_i = \mu_i.$$

The fundamental notions of kinematics, such as the position of a particle in space or its velocity, are generally regarded as observable or nontheoretical concepts. A proof that the concept of mass cannot be defined in terms of kinematical notions would therefore support the thesis of the theoreticity of the concept of mass. In order to study the logical relations among the fundamental notions of a theory, such as their logical independence, on the one hand, or their interdefinability, on the other, it is expedient, if not imperative, to axiomatize the theory and preferably to do it in such a way that the basic concepts under discussion are the primitive(undefined) notions in the axiomatized theory. As far as the concept of mass is concerned, there is hardly an axiomatization of classical particle mechanics that does not count this concept among its primitive notions.

In fact, as Gustav Kirchhoff's *Lectures on Mechanics*, or Heinrich Hertz's *Principles of Mechanics*, or more recently the axiomatic framework for classical particle mechanics proposed by Adonai Schlup Sant'Anna clearly show, even axiomatizations of mechanics that avoid the notion of force need the concept of mass as a primitive notion.

Any proof of the undefinability of mass in terms of other primitive notions can, of course, be given only within the framework of an axiomatization of mechanics. Let us choose for this purpose the widely known axiomatic formulation of classical particle mechanics proposed in 1953 by John Charles Chenoweth McKinsey and his collaborators, which is closely related to the axiomatization proposed by Patrick Suppes.

The axiomatization is based on five primitive notions: P, T, m, s, and f, where P and T are sets and m, s, and f are unary, binary, and ternary functions, respectively. The intended interpretation of P is a set of particles, denoted by p, that of T is a set of real numbers t measuring elapsed times(measured from some origin of time); the interpretation of the unary function m on P, i.e., $m(p)$, is the numerical value of the mass of particle p, while $s(p, t)$ is interpreted as the position vector of particle p at time t, and $f(p, t, i)$ as the ith force acting on particle p at time t, it being assumed that each particle is subjected to a number of different forces.

Clearly, A-6 is a formulation of Newton's second law of motion and, since for

$$\sum_{i-1}^{\infty} f(p,t,i) = 0$$

obviously $s(p,t) = a + bt$, A-6 also implies Newton's first law of motion. However, the question we are interested in is this: can it be rigorously demonstrated that the primitive m, which is intended to be interpreted as "mass," *cannot* be defined by means of the other primitive terms of the axiomatization, or at least not by means of the primitive notions that are used in the kinematical axioms? The standard procedure followed to prove that a given primitive of an axiomatization cannot be defined in terms of the other primitives of that axiomatization is the Padoa method, so called after the logician Alessandro Padoa, who invented it in 1900.

According to this method it is sufficient to find two interpretations of the axiomatic system that differ in the interpretation of the given primitive but retain the same interpretation for all the other primitives of the system.

For if the given primitive were to depend on the other primitives, the interpretation of the latter would uniquely determine the interpretation of the given primitive so that it would be impossible to find two interpretations as described.

Padoa's formulation of his undefinability proof has been criticized for not meeting all the requirements of logical rigor and, in particular, for its lack of a rigorous criterion for the "differentness" of interpretations. It has therefore been reformulated by, among others, John C. C. McKinsey, Evert Willem Beth, and Alfred Tarski.

A similar argument proves the logical independence of m in the axiomatization proposed by Suppes. These considerations seem to suggest that, quite generally, the concept of mass cannot be defined in terms of kinematical conceptions and, as such conceptions correspond to observational notions, mass is thus a theoretical term.

In 1977 Jon Dorling challenged the general validity of such a conclusion. Recalling that in many branches of mathematical physics theoretical terms, e.g., the vector potentials in classical or in quantum electrodynamics, have been successfully eliminated in favour of observational terms, Dorling claimed that the asserted uneliminability results only from the "idiosyncratic choice" of the observational primitives.

Referring to G.W. Mackey's above axiomatization in which the acceleration of each particle is given as a function of its position and the positions of the other particles and not, as in McKinsey's or Suppes's axiomatization, of time only, Dorling declared: "The claim that the usual theoretical primitives of classical particle mechanics cannot be eliminated in favour of observational primitives seems therefore not only not to have been established by Suppes's results, but to be definitely controverted in the case of more orthodox axiomatizations such as Mackey's." The issue raised by Dorling has been revived, though without any reference to him, by the following relatively recent development.

In 1993 Hans-J ürgen Schmidt offered a new axiomatization of classical particle mechanics intended to lead to an essentially universal concept of mass. He noted that in former axiomatizations the inertial mass m_k had usually been introduced as a coefficient connected with the acceleration a_k of the kth particle in such a way that the products $m_k a_k$ satisfy a certain condition that is not satisfied by the a_k alone. "If this condition determines the coefficients m_k uniquely—up to a common factor—" he declared, "we have got the clue for the definition of mass. This definition often works if the defining condition is taken simply as a special force law, but then one will arrive at different concepts of mass."

In order to avoid this deficiency Schmidt chose instead of a force-determining condition one that is equivalent to the existence of a Lagrangian. This choice involves the difficult task of solving the socalled "inverse problem of Lagrangian mechanics" to find a variational principle for a given differential equation. This problem was studied as early as 1886 by Hermann von Helmholtz and solved insofar as he found the conditions necessary for the existence of a function L such that a given set of equations $G_j = 0$ are the Euler-Lagrange equations of the variational principle

$$ä + -Ldt = 0.$$

Assisted by Peter Havas's 1957 study of the applicability of the Lagrange formalism, Schmidt, on the basis of a somewhat simplified solution of the inverse problem, was able to construct his axiomatization, which defines inertial mass in terms of accelerations. The five primitive terms of the axiomatization are the set M of space-time events, the differential structure D of M, the simultaneity relation ó on M, the set P of particles, and the set of possible motions of P, the last being bijective mappings or "charts" of M into the four-dimensional continuum R. Six axioms are postulated in terms of these primitives, none of which represents an equivalent to a force law.

The fact that these kinematical axioms lead to a satisfactory definition of mass is in striking contrast to the earlier axiomatizations for which it could be shown, for instance, by use of the Padoa method, that the dynamical concept of mass is indefinable in kinematical language. This apparent contradiction prompted Kamlah to distinguish between two kinds of axiomatic approaches to particle mechanics, differing in their epistemological positions, which he called factualism and potentialism.

According to factualist ontology, which, as Kamlah points out, was proclaimed most radically in Ludwig Wittgenstein's 1922 *Tractatus Logico-Philosophicus*, "there are certain facts in the world which may be described by a basic language for which the rules of predicate logic hold, especially the substitution rule, which makes this language an *extensional* one. The basic language has not to be an observational language." According to the ontology of potentialism "the world is a totality of *possible experiences*. Not all possible experiences actually happen."

By distinguishing between a factualist and a potentialist axiomatization Kamlah claims to resolve that contradiction as follows: The concept of acceleration a_k contained in Schmidt's potentialist kinematics can be "defined" operationally in the language of factualist kinematics. However, Kamlah adds, such determinations of the meaning of concepts are not proper definitions though being indispensable in physics, and therefore the acceleration function a_k is a theoretical concept in particle kinematics.

This theoretical concept seems to be powerful enough in combinations with [Schmidt's additional axioms] to supply us with an explicit definition of mass. This result seems to be surprising but does not contradict the well established theorem that mass is theoretical(not explicitly definable) in particle kinematics.

The thesis of the theoretical status of the concept of inertial mass—whether based on the argument of the alleged impossibility of defining this concept in a noncircular operational way or on the claim that it is implicitly defined by its presence in the laws of motion or in the axioms of mechanics—has been challenged by the proponents of protophysics.

The programme of protophysics, a doctrine that was developed by the Erlangen School of Constructivism but can be partially traced back to Pierre Duhem and Hugo Dingler, is the reconstruction of physics on prescientific constructive foundations with due consideration for the technical construction of the measuring instruments to be used in physics.

Protophysics insists on a rigorous compliance with what it calls the methodical order of the pragmatic dependence of operational procedures, in the sense that an operation O_2 is pragmatically dependent upon an operation O_1 if O_2 can be carried out successfully only after O_1 has previously been carried out successfully. In accordance with the three fundamental notions in physics—space, time, and mass— protophysicists distinguish among(constructive) geometry, chronometry, and hylometry, the last one, the protophysics of mass, having been subject to far less attention that the other two. Protophysicists have dealt with the concept of charge, often called the fourth fundamental notion of physics, to an even more limited degree.

Strictly speaking, the first to treat "mass" as a hylometrical conception was Bruno Th üring, who contended that the classical law of gravitation has to form part of the measure-theoretical a priori of empirical physics. However, this notion of mass was, of course, the concept of gravitational mass.

As far as inertial mass is concerned, the mathematician and philosopher Paul Lorenzen was probably the first to treat "mass" from the protophysical point of view. Lorenzen's starting point, as in Weyl's definition of mass, is an inelastic collision of two bodies with initial velocities u_1 and u_2, respectively, where the common velocity of the collision is u.

That it is technically possible("hinreichend gut") to eliminate friction can be tested by repeating the process with different u_1 and u_2 and checking that the ratio r of the velocity changes $u_1 - u$ and $u_2 - u$ is a constant. However, the absence of friction cannot be defined in terms of this constant, for were it verified in the reference frame of the earth it would not hold in a reference frame in accelerated motion relative to the earth.

If an inertial system is defined as the frame in which this constancy has been established, it is a technical-practical question whether the earth is an inertial system.

Foucault's pendulum shows that it is not. Lorenzen proposed therefore that the astronomical fundamental coordinate system S, relative to which the averaged rotational motion of the galaxies is zero, serves as the inertial system.

Any derivation from a constant *r* must then be regarded and explained as a "perturbation." This proposed purely kinematical definition of an inertial system is equivalent to defining such a system by means of the principle of conservation of momentum. The statement that numbers m_1 and m_2 can be assigned by this method to bodies as measures of their "mass" is then the Ramsey sentence for applying the momentum principle for collision processes in *S*.

A protophysical determination of inertial mass without any recourse to an inertial reference frame or to "free motion" has been proposed by Peter Janich. Janich employs what he calls a "rope balance"("Seilwaage"), a wheel encircled by a rope that has a body attached to each end. The whole device can be moved, for instance, on a horizontal(frictionless) plane in accelerated motion relative to an arbitrary reference frame.

As Janich points out, the facts that the rope is constant in length and taut and that the two end pieces beyond the wheel are parallel and of equal length can be verified geometrically. If these conditions are satisfied the two bodies are said to be "tractionally equal," a relation that can be proved to be an equivalence relation.

The transition from this classification measurement to a metric measurement is established by a definition of "homogeneous density": a body is homogeneously dense if any two parts of it, equal in volume, are tractionally equal, it being assumed, of course, that the equality of volume, as that of length before, has been defined in terms of protophysical geometry.

The ability to produce technically homogeneously dense bodies such as pure metals or homogeneous alloys is also assumed. Finally, the mass-ratio m_A / m_B of two arbitrary bodies *A* and *B* is defined by the volume ratio V_A / V_B of two bodies *B* and *C*, provided that *C* is tractionally equal to *A*, *D* is tractionally equal to *B*, and *C* and *D* are parts of a homogeneously dense body. Thus the metrics of mass is reduced to the metrics of volume and length.

By assigning logical priority to the notion of density over that of mass Janich, in a sense, "vindicated" Newton's definition of mass as the product of volume and density—but of course, unlike Newton, without conceiving density as a primitive concept.

On the basis of this definition and measurement of inertial mass, an inertial reference system can be defined as that reference frame relative to which, for example, the conservation of linear momentum in an inelastic collision holds by checking the validity of equation all the terms of which are now protophysically defined.

Kamlah has shown how Janich's rope balance, which can also be used for a comparative measurement of masses, is an example of the far-reaching applicability of D'Alembert's principle.

This does not mean, however, that Kamlah accepts the doctrine of protophysics. His criticism of the claim that the constructivist measurement-instructions cannot be experimentally invalidated without circularity, though directed primarily against the protophysics of time, applies equally well to the protophysics of mass.

Friedrich Steinle also criticized Janich's definition of mass on the grounds that it yields a new conception of mass and not a purged reconstruction of Newton's conception because for Newton "mass" and "weight," though proportional to one another, were two independent concepts, whereas, Steinle contends in Janich's reconstruction this proportionality is part of the definition.

It may also be that Janich's definition of the homogeneous density of a body can hardly be reconciled with the pragmatic programme of protophysics; for to verify that *any* two parts of the body, equal in volume, are also tractionally equal would demand an *infinite* number of technical operations.

In all the definitions of inertial mass discussed so far, whether they have been proposed by protophysicists, by operationalists, or by advocates of any other school of the philosophy of physics, one fact has been completely ignored or at least thought to be negligible. This is the inevitable interaction of a physical object—be it a macroscopic body or a microphysical particle—with its environment.(In what follows we shall sometimes use the term "particle" also in the sense of a body and call the environment the "medium" or the "field.")

Under normal conditions the medium is air. But even if the medium is what is usually called a "vacuum," physics tells us that it is not empty space. In prerelativistic physics a vacuum was thought to be permeated by the ether; in modern physics and in particular in its quantum field theories, this so-called vacuum is said to contain quanta of thermal radiation or "virtual particles" that may even have their origin in the particle itself. Nor should we forget that even in classical physics the notion of an absolute or ideal vacuum was merely an idealization never attainable experimentally.

Ingeneral, if aparticle is acteduponbyaforce F, itsacceleration a in the medium can be expected to be smaller than the hypothetical acceleration a_0 it would experience when moving in free space. However, if $a < a_0$ then the mass m, defined by F/a, is greater than the mass m_0, defined by F/a_0. This allows us to write

$$m = m' + \delta m,$$

where m denotes the experimentally observable or "effective" mass of the particle, m_0 its hypothetical or "bare" mass, and δm the increase in inertia owing to the interaction of the particle with the medium.

These observations may have some philosophical importance. Should it turn out that there is no way to determine m_0, i.e., the inertial behaviour of a physical object when it is not affected by an interaction with a field, it would go far toward supporting the thesis that the notion of inertial mass is a theoretical concept. Let us therefore discuss in some detail how such interactions complicate the definition of inertial mass and lead to different designations of this notion corresponding to the medium being considered.

Conceptually and mathematically the least complicated notion of this kind is the concept of "hydrodynamical mass."

Its history can be traced back to certain early nineteenth-century theories that treated the ether as a fluid, and in its more proper sense in the mechanics of fluids to Sir George Gabriel Stokes's extensive studies in this field. However, the term "hydrodynamical mass" was only given currency in 1953 by Sir Charles Galton Darwin, the grandson of the famous evolutionist Charles Robert Darwin.

In order to understand the definition of this concept let us consider the motion of a solid cylinder of radius r moving through an infinite in compressible fluid, say water or air, of density ρ, with constant velocity v. The kinetic energy of the fluid is

$$E^f_{\text{kin}} = \frac{1}{2}\pi\rho r^2 v^2$$

and its mass per unit thickness is $M' = \pi\rho r$. If M denotes the mass of the cylinder per unit thickness, then the total kinetic energy of the fluid and cylinder is clearly

$$E_{\text{kin}} = \frac{1}{2}(M + M')v^2;$$

and if F denotes the external force in the direction of the motion of the cylinder, which sustains the motion, then the rate at which F does work, being equal to the rate of increase in E_{kin}, is given by

$$Fv = dE_{\text{kim}} / dt = (M + M')v dv / dt.$$

This shows that the cylinder experiences a resistance to its motion equal to $M' dv/dt$ per unit thickness owing to the presence of the fluid. Comparison with Newton's second law suggests that $M + M'$ be called the "virtual mass" of the cylinder and the added mass M' the "hydrodynamical mass."

It can be shown to be quite generally true that every moving body in a fluid medium is affected by an added mass so that its virtual mass is $M + kM'$, where the coefficient k depends on the shape of the body and the nature of the medium. Clearly the notion of "hydrodynamic mass" poses no special problems because it is formulated entirely within the framework of classical mechanics.

Much more problematic is the case in which the medium is not a fluid in the mechanical sense of the term but an electromagnetic field whether of external origin or one produced by the particle itself if it is a charged particle such as the electron.

Theories about electromagnetic radiative reactions have generally been constructed on the basis of balancing the energy-momentum conservation. But the earliest theory that a moving charged body experiences a retardation owing to its own radiation, so that its inertial mass appears to increase, was proposed by the Scottish physicist Balfour Stewart on qualitative thermodynamical arguments. Since a rather detailed historical account of the concept of mass in classical electromagnetic theory has been given elsewhere, we shall confine ourselves here to the following very brief discussion.

Joseph John Thomson, who is usually credited with having discovered the electron, seems also to have been the first to write on the electromagnetic mass of a charged particle.

Working within the framework of James Clerk Maxwell's theory of the electromagnetic field, Thomson calculated the field produced by a spherical particle of radius r, which carries a charge e and moves with constant velocity v.

He found that the kinetic energy of the electromagnetic field produced by this charge— this field playing the role of the medium as described above—is given by the expression

$$E_{\mathrm{kin}}^{\mathrm{elm}} = ke^2 v^2 / 2rc^2 ,$$

where the coefficient k, of the order of unity, depends on how the charge e is distributed in, or on, the particle.

Although Thomson still regarded the increase in inertial mass as a phenomenon analogous to a solid moving through a perfect fluid, subsequent elaborations of the concept of electromagnetic mass, such as those carried out by Oliver Heaviside, George Francis Fitzgerald, and, in particular, by Hendrick Antoon Lorentz, suggested that this notion may well have important philosophical consequences. For, whereas the previous tendency had generally been to interpret electromagnetic processes as manifestations of mechanical interactions, the new conception of electromagnetic mass seemed to clear the way toward a reversal of this logical order, i.e., to deduce mechanics from the laws of electromagnetism.

If successful, such a theory would explain all processes in nature in terms of convection currents and their electromagnetic radiation, stripping the "stuff" of the world of its material substantiality. However, such an electromagnetic world-picture could be established only if it could be proved that m_0, the mechanical or bare mass of a charged particle, has no real existence.

Walter Kaufmann, whose well known experiments on the velocity dependence of inertial mass played an important role in these deliberations, claimed in 1902 that m_0, which he called the "real mass"("wirkliche Masse")—in contrast to m_{elm}, which he called the "apparent mass"("scheinbare Masse")—is zero, so that "the total mass of the electron is merely an electromagnetic phenomenon."

At the same time, Max Abraham, in a study that can be regarded as the first field-theoretic treatment of elementary particles, showed that, strictly speaking, the electromagnetic mass is not a scalar but rather a tensor with the symmetry of an ellipsoid of revolution and proclaimed: "The inertia of the electron originates in the electromagnetic field." However, he took issue with Kaufmann's terminology, for, as he put it, "the often used terms of 'apparent' and 'real' mass lead to confusion. For the 'apparent' mass, in the mechanical sense, is real, and the 'real' mass is apparently unreal."

Lorentz, the revered authority in this field, was more reserved. In a talk "On the Apparent Mass of Ions," as he used to call charged particles, he declared in 1901: "The question of whether the ion possesses in addition to its apparent mass also a real mass is of extraordinary importance; for it touches upon the problem of the connection between ponderable matter and the ether and electricity; I am far from being able to give a decisive answer."

Furthermore, in his lectures at Columbia University in 1906 he even admitted: "After all, by our negation of the existence of material mass, the negative electron has lost much of its substantiality.

We must make it preserve just so much of it that we can speak of forces acting on its parts, and that we can consider it as maintaining its form and magnitude. This must be regarded as an inherent property, in virtue of which the parts of the electron cannot be torn asunder by the electric forces acting on them(or by their mutual repulsion, as we may say)."

It should be recalled that at the same time Henri Poincaré also insisted on the necessity of ascribing nonelectromagnetic stresses to the electron in order to preserve the internal stability of its finite charge distribution. But clearly, such a stratagem would put an end to the theory of a purely electromagnetic nature of inertial mass.

The only way to save it would have been to describe the electron as a structureless point charge, which means to take $r = 0$. But then, as can be seen from equation, the energy of the self-interaction and thus the mass of the electron would become infinite.

Classical electromagnetic theory has never resolved this problem. As we shall see in what follows, the same problem of a divergence to infinity also had to be faced by the modern field theory of quantum electrodynamics.

With the advent of the special theory of relativity in the early years of the twentieth century, physicists and philosophers focused their attention on the concept of relativistic mass. Since this notion will be dealt with in the following chapter we shall turn immediately to the quantum mechanical treatment of inertial mass but for the time being only insofar as the medium affecting the mass of a particle consists of other particles arranged in a periodic crystal structure.

This is a subject studied in the quantum theory of solids or condensed matter and leads to the notion of effective mass. More specifically, we consider the case of an electron moving under the influence of an external force F through a crystal.

Obviously, m has a constant value only for energy bands of the form $E = E_0 \pm$ const. k.

But even in this case the effective mass may differ from the value of the inertial mass of a free electron. This difference is, of course, to be expected; for in general the acceleration of an electron moving under a given force in a crystal may well differ from the acceleration of an electron that is moving under the same force in free space. What is more difficult to understand intuitively is the fact that, owing to reflections by the crystal lattice, an electron can move in a crystal in the direction opposite to that it would have in free space. In this case the effective mass m^* is negative.

We conclude this survey with a brief discussion of the concepts of bare mass and experimental or observed mass as they are used in quantum electrodynamics, which, like every field theory, ascribes a field aspect to particles and all other physical entities and studies, in particular, the interactions of electrons with the electromagnetic field or its quanta, the photons. Soon after the birth of quantum mechanics it became clear that a consistent treatment of the problems of emission, absorption, and scattering of electromagnetic radiation requires the quantization of the electromagnetic field. In fact, Planck's analysis of the spectral distribution of blackbody radiation, which is generally hailed as having inaugurated quantum theory, is, strictly speaking, a subject of quantum electrodynamics.

Although no other physical theory has ever achieved such spectacular agreement between theoretical predictions and experimental measurements, some physicists, including Paul A. M. Dirac himself, have viewed it with suspicion because of its use of the so-called "renormalization" procedure, which was designed to cope with the divergences of self energy or mass, a problem that, as noted above, was left unresolved by classical electromagnetic theory.

It reappeared in quantum electrodynamics for the first time in 1930 in J. Robert Oppenheimer's calculation of the interaction between the quantum electromagnetic field and an atomic electron. "It appears improbable," said Oppenheimer, "that the difficulties discussed in this work will be soluble without an adequate theory of the masses of electron and proton, nor is it certain that such a theory will be possible on the basis of the special theory of relativity." The "adequate theory" envisaged by Oppenheimer took about twenty years to reach maturity.

As is well known, in modern field theory a particle such as an electron constantly emits and reabsorbs virtual particles such as photons. The application of quantum-mechanical perturbation theory to such a process leads to an infinite result for the self-energy or mass of the electron.(Technically speaking, such divergences are the consequences of the pointlike nature of the "vertex" in the Feynman diagram of the process.) Here it is, of course, this "cloud" of virtual photons that plays the role of the medium in the sense discussed above.

As early as the first years of the 1940s, Hendrik A. Kramers, the longtime collaborator of Niels Bohr, suggested attacking this problem by sharply distinguishing between what he called mechanical mass, as used in the Hamiltonian, and observable mass; but it was only in the wake of the famous four-day Shelter Island Conference of June 1947 that a way was found to resolve—or perhaps only to circumvent—the divergences of mass in quantum electrodynamics.

Perhaps inspired by Kramers's remarks at the conference, Hans Albrecht Bethe realized immediately— actually during his train ride back from Shelter Island—that the Lamb shift can be accounted for by quantum electrodynamics if this theory is appropriately interpreted. He reasoned that when calculating the self-energy correction for the emission and reabsorption of a photon by a bound electron, the divergent part of the energy shift can be identified with the self-mass of the electron.

Hence, in the calculation of the energy difference for the bound-state levels, as in the Lamb shift, the energy shift remains finite since both levels contain the same, albeit infinite, self-mass terms that cancel each other out in the subtraction.

It is this kind of elimination of infinities, based on the impossibility of measuring the bare mass m_0 by any conceivable experiment that constitutes the renormalization of mass in quantum electrodynamics. A more detailed exposition of the physics of mass renormalization can be found in standard texts on quantum field theory, and its mathematical features in John Collin's treatise.

The reader interested in the historical aspects of the subject is referred to the works of Olivier Darrigol and Seiya Aramaki, and the philosopher of contemporary physics to the essays by Paul Teller.

Chapter 9

Classical Electromagnetism

Classical electromagnetism(or classical electrodynamics) is a theory of electromagnetism that was developed over the course of the 19th century, most prominently by James Clerk Maxwell. It provides an excellent description of electromagnetic phenomena whenever the relevant length scales and field strengths are large enough that quantum mechanical effects are negligible. Maxwell's equations and the Lorentz force law form the basis for the theory of classical electromagnetism.

There is, however, a scalar function called the electrical potential that can help. Unfortunately, this definition has a caveat. From Maxwell's equations, it is clear that is not always zero, and hence the scalar potential alone is insufficient to define the electric field exactly. As a result, one must resort to adding a correction factor, which is generally done by subtracting the time derivative of the A vector potential described below.

Whenever the charges are quasistatic, however, this condition will be essentially met, so there will be few problems.(As a side note, by using the appropriate gauge transformations, one can define V to be zero and define E entirely as the negative time derivative of A, however, this is rarely done because a) it's a hassle and more important, b) it no longer satisfies the requirements of the Lorenz gauge and hence is no longer relativistically invariant).

ELECTROMAGNETIC WAVES

A changing electromagnetic field propagates away from its origin in the form of a wave. These waves travel in vacuum at the speed of light and exist in a wide spectrum of wavelengths. Examples of the dynamic fields of electromagnetic radiation(in order of increasing frequency): radio waves, microwaves, light(infrared, visible light and ultraviolet), x-rays and gamma rays. In the field of particle physics this electromagnetic radiation is the manifestation of the electromagnetic interaction between charged particles.

Electromagnetic radiation(sometimes abbreviated EMR) takes the form of self-propagating waves in a vacuum or in matter. EM radiation has an electric and magnetic field component which oscillate in phase perpendicular to each other and to the direction of energy propagation. Electromagnetic radiation is classified into types according to the frequency of the wave, these types include(in order of increasing frequency): radio waves, microwaves, terahertz radiation, infrared radiation, visible light, ultraviolet radiation, X-rays and gamma rays. Of these, radio waves have the longest wavelengths and Gamma rays have the shortest.

A small window of frequencies, called visible spectrum or light, is sensed by the eye of various organisms, with variations of the limits of this narrow spectrum.

EM radiation carries energy and momentum, which may be imparted when it interacts with matter.

THEORY

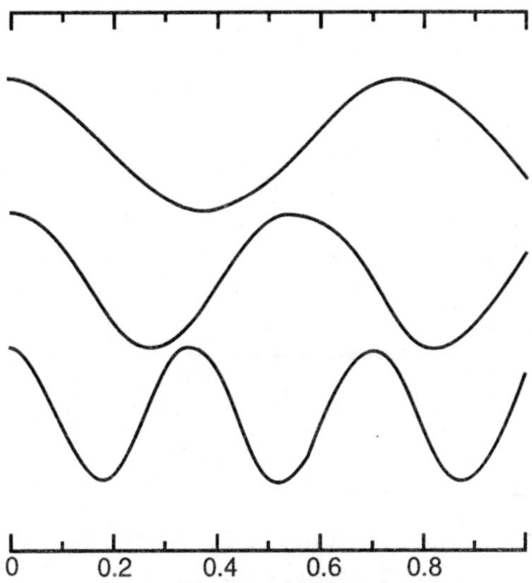

Fig. Shows three Electromagnetic Modes(Blue, Green and Red with a Distance Scale in Microns along the x-axis.

Electromagnetic waves were first postulated by James Clerk Maxwell and subsequently confirmed by Heinrich Hertz. Maxwell derived a wave form of the electric and magnetic equations, revealing the wave-like nature of electric and magnetic fields, and their symmetry. Because the speed of EM waves predicted by the wave equation coincided with the measured speed of light, Maxwell concluded that light itself is an EM wave. According to Maxwell's equations, a time-varying electric field generates a magnetic field and *vice versa.*

Therefore, as an oscillating electric field generates an oscillating magnetic field, the magnetic field in turn generates an oscillating electric field, and so on. These oscillating fields together form an electromagnetic wave. A quantum theory of the interaction between electromagnetic radiation and matter such as electrons is described by the theory of quantum electrodynamics.

PROPERTIES

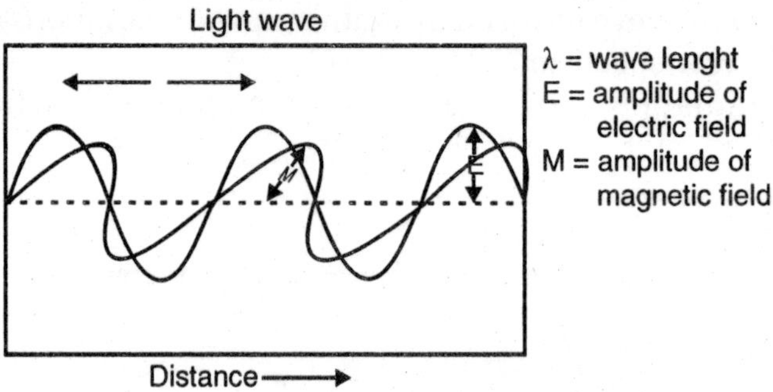

Electromagnetic waves can be imagined as a self-propagating transverse oscillating wave of electric and magnetic fields. This diagram shows a plane linearly polarized wave propagating from right to left. The electric field is in a vertical plane, the magnetic field in a horizontal plane.

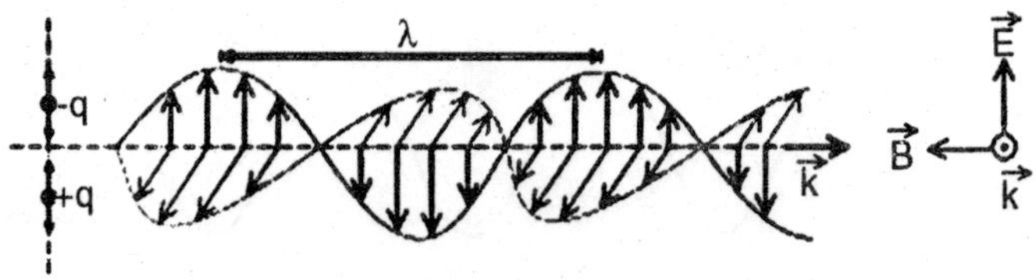

Electric and magnetic fields obey the properties of superposition, so fields due to particular particles or time-varying electric or magnetic fields contribute to the fields due to other causes.(As these fields are vector fields, all magnetic and electric field vectors add together according to vector addition.) These properties cause various phenomena including refraction and diffraction. For instance, a travelling EM wave incident on an atomic structure induces oscillation in the atoms, thereby causing them to emit their own EM waves. These emissions then alter the impinging wave through interference.

Since light is an oscillation, it is not affected by travelling through static electric or magnetic fields in a linear medium such as a vacuum.

In nonlinear media such as some crystals, however, interactions can occur between light and static electric and magnetic fields - these interactions include the Faraday effect and the Kerr effect.

In refraction, a wave crossing from one medium to another of different density alters its speed and direction upon entering the new medium. The ratio of the refractive indices of the media determines the degree of refraction, and is summarized by Snell's law. Light disperses into a visible spectrum as light is shone through a prism because of the wavelength dependant refractive index of the prism material.

The physics of electromagnetic radiation is electrodynamics, a subfield of electromagnetism. EM radiation exhibits both wave properties and particle properties at the same time. The wave characteristics are more apparent when EM radiation is measured over relatively large timescales and over large distances, and the particle characteristics are more evident when measuring small distances and timescales.

Both characteristics have been confirmed in a large number of experiments. There are experiments in which the wave and particle natures of electromagnetic waves appear in the same experiment, such as the diffraction of a single photon. When a single photon is sent through two slits, it passes through both of them interfering with itself, as waves do, yet is detected by a photomultiplier or other sensitive detector only once. Similar self-interference is observed when a single photon is sent into a Michelson interferometer or other interferometers.

WAVE MODEL

An important aspect of the nature of light is frequency. The frequency of a wave is its rate of oscillation and is measured in hertz, the SI unit of frequency, where one hertz is equal to one oscillation per second. Light usually has a spectrum of frequencies which sum together to form the resultant wave. Different frequencies undergo different angles of refraction.

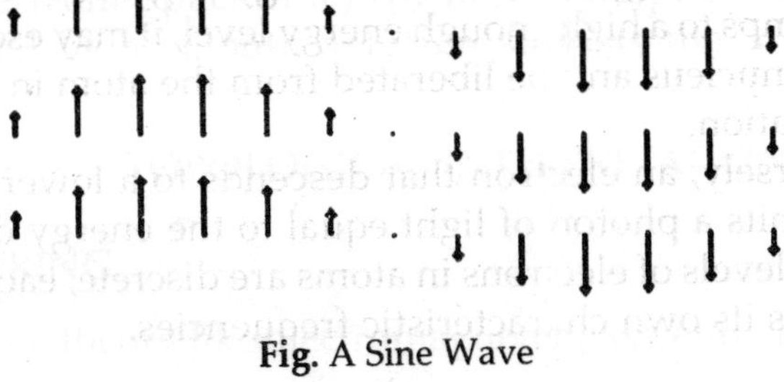

Fig. A Sine Wave

A wave consists of successive troughs and crests, and the distance between two adjacent crests or troughs is called the wavelength. Waves of the electromagnetic spectrum vary in size, from very long radio waves the size of buildings to very short gamma rays smaller than atom nuclei. Frequency is inversely proportional to wavelength, according to the equation:

$$u = f\lambda$$

where v is the speed of the wave(c in a vacuum, or less in other media), f is the frequency and ë is the wavelength. As waves cross boundaries between different media, their speeds change but their frequencies remain constant.

Interference is the superposition of two or more waves resulting in a new wave pattern. If the fields have components in the same direction, they constructively interfere, while opposite directions cause destructive interference. The energy in electromagnetic waves is sometimes called radiant energy.

PARTICLE MODEL

Because energy of an EM wave is quantized, in the particle model of EM radiation, a wave consists of discrete packets of energy, or quanta, called photons. The frequency of the wave is proportional to the magnitude of the particle's energy. Moreover, because photons are emitted and absorbed by charged particles, they act as transporters of energy. The energy per photon can be calculated by Planck's equation:

$$E = hf$$

where E is the energy, h is Planck's constant, and f is frequency. This photon-energy expression is a particular case of the energy levels of the more general *electromagnetic oscillator* whose average energy, which is used to obtain Planck's radiation law, can be shown to differ sharply from that predicted by the equipartition principle at low temperature, thereby establishes a failure of equipartition due to quantum effects at low temperature.

As a photon is absorbed by an atom, it excites an electron, elevating it to a higher energy level. If the energy is great enough, so that the electron jumps to a high enough energy level, it may escape the positive pull of the nucleus and be liberated from the atom in a process called photoionisation.

Conversely, an electron that descends to a lower energy level in an atom emits a photon of light equal to the energy difference. Since the energy levels of electrons in atoms are discrete, each element emits and absorbs its own characteristic frequencies.

Together, these effects explain the absorption spectra of light. The dark bands in the spectrum are due to the atoms in the intervening medium absorbing different frequencies of the light. The composition of the medium through which the light travels determines the nature of the absorption spectrum. For instance, dark bands in the light emitted by a distant star are due to the atoms in the star's atmosphere. These bands correspond to the allowed energy levels in the atoms. A similar phenomenon occurs for emission.

As the electrons descend to lower energy levels, a spectrum is emitted that represents the jumps between the energy levels of the electrons. This is manifested in the emission spectrum of nebulae. Today, scientists use this phenomenon to observe what elements a certain star is composed of. It is also used in the determination of the distance of a star, using the so-called red shift.

SPEED OF PROPAGATION

Any electric charge which accelerates, or any changing magnetic field, produces electromagnetic radiation. Electromagnetic information about the charge travels at the speed of light.

Accurate treatment thus incorporates a concept known as retarded time(as opposed to advanced time, which is unphysical in light of causality), which adds to the expressions for the electrodynamic electric field and magnetic field.

These extra terms are responsible for electromagnetic radiation. When any wire(or other conducting object such as an antenna) conducts alternating current, electromagnetic radiation is propagated at the same frequency as the electric current.

Depending on the circumstances, it may behave as a wave or as particles. As a wave, it is characterized by a velocity(the speed of light), wavelength, and frequency.

When considered as particles, they are known as photons, and each has an energy related to the frequency of the wave given by Planck's relation $E = hí$, where E is the energy of the photon,

$$h = 6.626 \times 10^{-} \text{ J·s}$$

is Planck's constant, and $í$ is the frequency of the wave. One rule is always obeyed regardless of the circumstances: EM radiation in a vacuum always travels at the speed of light, *relative to the observer*, regardless of the observer's velocity.(This observation led to Albert Einstein's development of the theory of special relativity.) In a medium(other than vacuum), velocity factor or refractive index are considered, depending on frequency and application. Both of these are ratios of the speed in a medium to speed in a vacuum.

ELECTROMAGNETIC SPECTRUM

Legend:
ã = Gamma rays
HX = Hard X-rays
SX = Soft X-Rays
EUV = Extreme ultraviolet
NUV = Near ultraviolet
Visible light
NIR = Near infrared
MIR = Moderate infrared
FIR = Far infrared
Radio waves:
EHF = Extremely high frequency(Microwaves)
SHF = Super high frequency(Microwaves)
UHF = Ultrahigh frequency(Microwaves)
VHF = Very high frequency
HF = High frequency
MF = Medium frequency
LF = Low frequency
VLF = Very low frequency
VF = Voice frequency
ELF = Extremely low frequency

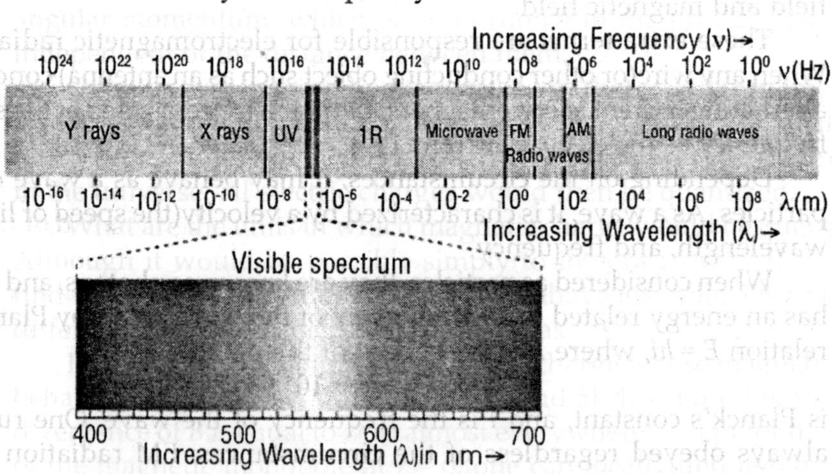

Fig. Electromagnetic Spectrum with Light Highlighted

Generally, EM radiation is classified by wavelength into electrical energy, radio, microwave, infrared, the visible region we perceive as light, ultraviolet, X-rays and gamma rays. The behaviour of EM radiation depends on its wavelength.

Higher frequencies have shorter wavelengths, and lower frequencies have longer wavelengths. When EM radiation interacts with single atoms and molecules, its behaviour depends on the amount of energy per quantum it carries. Spectroscopy can detect a much wider region of the EM spectrum than the visible range of 400 nm to 700 nm. A common laboratory spectroscope can detect wavelengths from 2 nm to 2500 nm. Detailed information about the physical properties of objects, gases, or even stars can be obtained from this type of device. It is widely used in astrophysics. For example, hydrogen atoms emit radio waves of wavelength 21.12 cm.

CLASS	FREQUENCY	WAVELENGTH	ENERGY
	300 EHz	1 pm	1.24 MeV
γ			
	30 EHz	10 pm	124 keV
HX			
	3 EHz	100 pm	12.4 keV
SX			
	300 PHz	1 nm	1.24 keV
	30 PHz	10 nm	124 eV
EUV			
	3 PHz	100 nm	12.4 eV
NUV			
	300 THz	1 μm	1.24 eV
NIR			
	30 THz	10 μm	124 meV
MIR			
	3 THz	100 μm	12.4 meV
FIR			
	300 GHz	1 mm	1.24 meV
EHF			
	30 GHz	1 cm	124 μeV
SHF			
	3 GHz	1 dm	12.4 μeV
UHF			
	300 MHz	1 m	1.24 μeV
VHF			
	30 MHz	1 dam	124 neV
HF			
	3 MHz	1 hm	12.4 neV
MF			
	300 kHz	1 km	1.24 neV
LF			
	30 kHz	10 km	124 peV
VLF			
	3 kHz	100 km	12.4 peV
VF			
	300 Hz	1 Mm	1.24 peV
ELF			
	30 Hz	10 Mm	124 feV

LIGHT

EM radiation with a wavelength between approximately 400 nm and 700 nm is detected by the human eye and perceived as visible light. Other wavelengths, especially nearby infrared(longer than 700 nm) and ultraviolet (shorter than 400 nm) are also sometimes referred to as light, especially when the visibility to humans is not relevant. If radiation having a frequency in the visible region of the EM spectrum reflects off of an object, say, a bowl of fruit, and then strikes our eyes, this results in our visual perception of the scene.

Our brain's visual system processes the multitude of reflected frequencies into different shades and hues, and through this not-entirely-understood psychophysical phenomenon, most people perceive a bowl of fruit.

At most wavelengths, however, the information carried by electromagnetic radiation is not directly detected by human senses. Natural sources produce EM radiation across the spectrum, and our technology can also manipulate a broad range of wavelengths. Optical fibre transmits light which, although not suitable for direct viewing, can carry data that can be translated into sound or an image. The coding used in such data is similar to that used with radio waves.

RADIO WAVES

Radio waves can be made to carry information by varying a combination of the amplitude, frequency and phase of the wave within a frequency band. When EM radiation impinges upon a conductor, it couples to the conductor, travels along it, and induces an electric current on the surface of that conductor by exciting the electrons of the conducting material. This effect(the skin effect) is used in antennas. EM radiation may also cause certain molecules to absorb energy and thus to heat up; this is exploited in microwave ovens.

DERIVATION

Electromagnetic waves as a general phenomenon were predicted by the classical laws of electricity and magnetism, known as Maxwell's equations. If you inspect Maxwell's equations without sources(charges or currents) then you will find that, along with the possibility of nothing happening, the theory will also admit nontrivial solutions of changing electric and magnetic fields. Beginning with Maxwell's equations for free space:

$$\Delta \cdot E = 0$$

$$\Delta \times E = -\frac{\partial}{\partial t} B$$

$$\Delta \cdot B = 0$$

$$\Delta \times B = \mu_0 \epsilon_0 \frac{\partial}{\partial t} E$$

where Δ is a vector differential operator.

One solution,

$$E = B = 0$$

is trivial.

To see the more interesting one, we utilize vector identities, which work for any vector, as follows:

$$\Delta \times (\Delta \times A) = \Delta(\Delta \cdot A) - \Delta^2 A$$

To see how we can use this take the curl of equation(2):

$$\Delta \times (\Delta \times E) = \Delta \times \left(-\frac{\partial B}{\partial t} \right)$$

Evaluating the left hand side:

$$\Delta \times (\Delta \times E) = \Delta(\Delta \cdot E) - \Delta^2 E = -\Delta^2 E$$

where we simplified the above by using equation(1).

Evaluate the right hand side:

$$\Delta \times \left(-\frac{\partial B}{\partial t} \right) = -\frac{\partial}{\partial t}(\Delta \times B) = -\mu_0 \in_0 \frac{\partial^2}{\partial t^2} E$$

Equations(6) and(7) are equal, so this results in a vector-valued differential equation for the electric field, namely

$$\nabla^2 E = \mu_0 \in_0 \frac{\partial^2}{\partial t^2} E$$

Applying a similar pattern results in similar differential equation for the magnetic field:

$$\nabla^2 B = \mu_0 \in_0 \frac{\partial^2}{\partial t^2} E$$

These differential equations are equivalent to the wave equation:

$$\nabla^2 f = \frac{1}{c_0^2} \frac{\partial^2 f}{\partial t^2}$$

where c_0 is the speed of the wave in free space and f describes a displacement

Notice that in the case of the electric and magnetic fields, the speed is:

$$c_0 = \frac{1}{\sqrt{\mu_0 \in_0}}$$

Which, as it turns out, is the speed of light in free space? Maxwell's equations have unified the permittivity of free space ε_0, the permeability of free space μ_0, and the speed of light itself, c_0. Before this derivation it was not known that there was such a strong relationship between light and electricity and magnetism.

But these are only two equations and we started with four, so there is still more information pertaining to these waves hidden within Maxwell's equations. Let's consider a generic vector wave for the electric field.

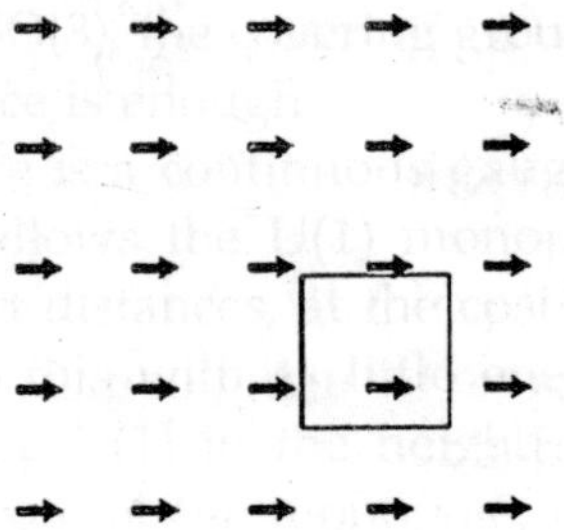

Fig. The Field $\hat{x}$.

$$E = E_0 f(\hat{k}\cdot x - c_0 t)$$

Here E_0 is the constant amplitude, f is any second differentiable function, $\hat{k}$ is a unit vector in the direction of propagation, and is a position vector. We observe that $f(\hat{k}\cdot x - c_0 t)$ is a generic solution to the wave equation. In other words

$$\nabla^2 f(\hat{k}\cdot x - c_0 t) = \frac{1}{co^2}\frac{\partial^2}{\partial^2 t} f(\hat{k}\cdot x - c_0 t),$$

for a generic wave traveling in the $\hat{k}$ direction.

This form will satisfy the wave equation, but will it satisfy all of Maxwell's equations, and with what corresponding magnetic field?

$$\nabla\cdot E = \hat{k}\cdot E_0 f'(\hat{k}\cdot x - c_0 t) = 0$$

$$E\cdot\hat{k} = 0$$

The first of Maxwell's equations implies that electric field is orthogonal to the direction the wave propagates.

$$\nabla\times E = \hat{k}\times E_0 f'(\hat{k}\cdot x - c_0 t) = -\frac{\partial}{\partial t}B$$

$$B = \frac{1}{c_0}\hat{k}\times E$$

The second of Maxwell's equations yields the magnetic field. The remaining equations will be satisfied by this choice of E,B. Not only are the electric and magnetic field waves traveling at the speed of light, but they have a special restricted orientation and proportional magnitudes, $E_0 = c_0 B_0$, which can be seen immediately from the Poynting vector. The electric field, magnetic field, and direction of wave propagation are all orthogonal, and the wave propagates in the same direction as $E \times B$.

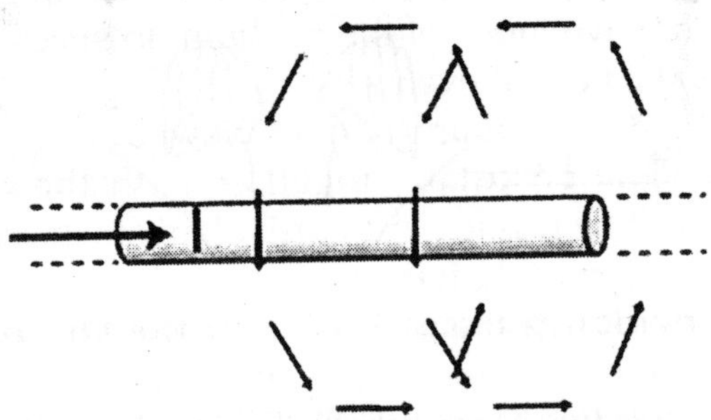

Fig. The Magnetic Field of a Long, Straight Wire.

From the viewpoint of an electromagnetic wave traveling forward, the electric field might be oscillating up and down, while the magnetic field oscillates right and left; but this picture can be rotated with the electric field oscillating right and left and the magnetic field oscillating down and up.

This is a different solution that is traveling in the same direction. This arbitrariness in the orientation with respect to propagation direction is known as polarization.

SPEED OF LIGHT

"Lightspeed" redirects here. For other uses, see Lightspeed (disambiguation). The speed of light in the vacuum of free space is an important physical constant usually denoted by the symbol c_0 or simply c. The metre is defined such that the speed of light in free space is *exactly* 299,792,458 metres per second (m/s).

The speed of light is of fundamental importance in physics. It is the speed of not just visible light, but of all electromagnetic radiation, as well as gravitational waves and anything having zero rest mass. In Einstein's theory of relativity the speed of light plays the crucial role of a conversion factor between space and time within spacetime. This theory together with the principle of causality requires that no matter or information can travel faster than the speed of light.

The speed of light is so great that for many purposes it can be regarded as infinite. However, where long distances or accurate time measurements are involved, the finite speed of light can be important. For example, in the Global Positioning System (GPS), a GPS receiver measures its distance to satellites based on how long it takes for a radio signal to arrive from the satellite. In astronomy, distances are often measured in light-years, the distance light travels in a year (around ten trillion kilometers). The speed of light when it passes through a transparent or translucent material medium, like glass or air, is less than its speed in a vacuum. The speed is controlled by the refractive index of the medium. In specially-prepared media, the speed can be tiny, or even zero.

The speed of light in vacuum is now viewed as a fundamental physical constant. This postulate, together with the principle of relativity that all inertial frames are equivalent, forms the basis of Einstein's theory of special relativity.

Experimental evidence has shown that the speed of light is independent of the motion of the source. It has also been confirmed experimentally that the two-way speed of light (for example from a source, to a mirror, and back again) is constant. It is not, however, possible to measure the one-way speed of light (for example from a source to a distant detector) without some convention as to how clocks at the source and receiver should be synchronized. Einstein (who was aware of this fact) postulated that the speed of light should be taken as constant in all cases, one-way and two-way.

An observer moving with respect to a collection of light sources would find that light from the sources ahead would be blueshifted while light from those behind was redshifted.

Use of the Symbol c' for the Speed of Light

The symbol *c* for 'constant' or the Latin *celeritas* ("swiftness") is used for the speed of light in free space, and in this article *c* is used exclusively this way. However, some authors use *c* for the speed of light in material media. To avoid confusion, and for consistency with other constants of free space such as $\grave{\imath}_0$, $\mathring{a}_0$ and Z_0, international bodies such as the International Bureau of Weights and Measures(BIPM) recommend using c_0 for the speed of light in free space.

In branches of physics in which the speed of light plays an important part, such as in relativity, it is common to use a system of units known as "natural units" in which *c* is 1; thus no symbol for the speed of light is required.

Causality and Information Transfer

According to the theory of special relativity, causality would be violated if information could travel faster than *c* in one reference frame. In some other reference frames, the information would be received before it had been sent, so the "effect" could be observed before the "cause". Such a violation of causality has never been recorded.

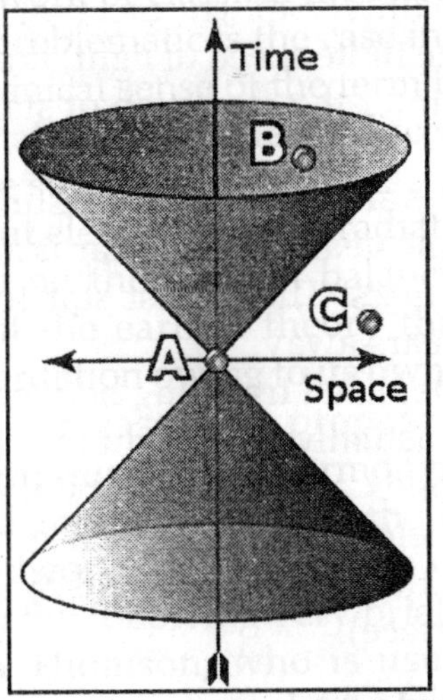

Fig. A Light Cone Defines Locations that are in Causal Contact and Those that are not.

To put it another way, information propagates to and from a point from regions defined by a light cone. The interval AB in the diagram to the right is "time-like" (that is, there is a frame of reference in which event A and event B occur at the same location in space, separated only by their occurring at different times, and if A precedes B in that frame then A precedes B in all frames: there is no frame of reference in which event A and event B occur simultaneously). Thus, it is hypothetically possible for matter (or information) to travel from A to B, so there can be a causal relationship (with A the "cause" and B the "effect").

On the other hand, the interval AC in the diagram to the right is "space-like" (that is, there is a frame of reference in which event A and event C occur simultaneously, separated only in space; see simultaneity). However, there are also frames in which A precedes C or in which C precedes A. Barring some way of traveling faster than light, it is not possible for any matter (or information) to travel from A to C or from C to A. Thus there is no causal connection between A and C.

Speed of Light in Astronomy

The speed of light is particularly important in astronomy. Due to the vast distances involved it can take a very long time for light to travel from its source to Earth. For example, it takes 13 billion years for light to travel to Earth from the faraway galaxies viewed in the Hubble Ultra Deep Field images. Those photographs, taken today, capture images of the galaxies as they appeared 13 billion years ago (near the beginning of the universe). The fact that farther-away objects appear younger (due to the finite speed of light) is crucial in astronomy, allowing astronomers to infer the evolution of stars, galaxies, and the universe itself.

Astronomical distances are sometimes measured in light-years, the distance light travels in one year. A light year is around 9 trillion km, 6 trillion miles, or 0.3 parsecs. The closest star to Earth (besides the sun) is around 4.2 light years away.

The beam of light is depicted traveling between the Earth and the Moon in the same time it actually takes light to scale the real distance between them: 1.255 seconds at its mean orbital distance (surface to surface). The light beam helps provide the sense of scale of the Earth-Moon system relative to the Sun, which is 8.28 light-minutes away (photosphere to Earth surface).

COMMUNICATIONS AND GPS

The speed of light is of relevance to communications. For example, given the equatorial circumference of the Earth is about 40,075 km and c about 300,000 km/s, the theoretical shortest time for a piece of information to travel half the globe along the surface is 0.066838 s.

When light is traveling around the globe in an optical fiber, the actual transit time is longer, in part because the speed of light is slower by about 35% in an optical fiber, depending on its refractive index n, $v = c/n$. Furthermore, straight lines rarely occur in global communications situations, and delays are created when the signal passes through an electronic switch or signal regenerator. A typical time as of 2004 for a U.S. to Australia or Japan computer-to-computer ping is 0.18 s. The speed of light additionally affects wireless communications design.

Another consequence of the finite speed of light is that communications between the Earth and spacecraft are not instantaneous. There is a brief delay from the source to the receiver, which becomes more noticeable as distances increase.

This delay was significant for communications between ground control and Apollo 8 when it became the first spacecraft to orbit the Moon: for every question, the ground control station had to wait at least three seconds for the answer to arrive.

The communications delay between Earth and Mars is almost ten minutes. As a consequence of this, if a robot on the surface of Mars were to encounter a problem, its human controllers would not be aware of it until ten minutes later; it would then take at least a further ten minutes for instructions to travel from Earth to Mars.

This effect forms the basis of the Global Positioning System (GPS) and similar navigation systems. A position on Earth can be determined by means of the delays in radio signals received from a number of satellites, each carrying a very accurate atomic clock, and very carefully synchronized.

To work properly, this method requires that (among many other effects) the relative motion of satellite and receiver be taken into effect, which was how (on an interplanetary scale) the finite speed of light was originally discovered (see the following section).

The speed of light can also be of concern over very short distances. In supercomputers, the speed of light imposes a limit on how quickly data can be sent between processors.

If a processor operates at 1 GHz, a signal can only travel a maximum of 300 mm (about one foot) in a single cycle. Processors must therefore be placed close to each other to minimize communication latencies, which can cause difficulty with cooling. If clock frequencies continue to increase, the speed of light will eventually become a limiting factor for the internal design of single chips.

Constant Velocity from All Inertial Reference Frames

Most individuals are accustomed to the addition rule of velocities: if two cars approach each other from opposite directions, each traveling at a speed of 50 km/h, relative to the road surface, one expects that each car will measure the other as approaching at a combined speed of 50 + 50 = 100 km/h to a very high degree of accuracy. However, as speeds increase this rule becomes less accurate.

Two spaceships approaching each other, each traveling at 90% the speed of light relative to some third observer, would not measure each other as approaching at 90% + 90% = 180% the speed of light; instead they each measure the other as approaching at slightly less than 99.5% the speed of light. This last result is given by the Einstein velocity-addition formula:

$$u = \frac{v_1 + v_2}{1 + (v_1 . v_2) / c^2}$$

where v_1 and v_2 are the velocities of the spaceships as measured by the third observer, and u is the measured velocity of either space ship as observed by the other. This reduces to {{{1}}} for sufficiently small values of v_1 and v_2 (such as those typically encountered in common daily experiences), as the term $(v_1 \cdot v_2)/c^2$ approaches zero, reducing the denominator to 1.

If one of the velocities for the above formula (or both) are c, the final result is c, as is expected if the speed of light is the same in all reference frames. Another important result is that this formula always returns a value which is less than c whenever v_1 and v_2 are less than c: this shows that no acceleration in any frame of reference can cause one to exceed the speed of light with respect to another observer. Thus c acts as a speed limit for all objects with respect to all other objects in special relativity.

Luminiferous Aether (discredited)

Before the advent of special relativity, it was believed that light traveled through a medium called the luminiferous aether. Maxwell's equations predict a given speed of light, in much the same way as is the speed of sound in air. The speed of sound in air is relative to the movement of the air itself, and the speed of sound in air with respect to an observer may be changed if the observer is moving with respect to the air (or vice versa). The speed of light was believed to be relative to a medium of transmission for light that acted as air does for the transmission of sound—the luminiferous aether.

The Michelson–Morley experiment, arguably the most famous and useful null-result experiment in the history of physics, was designed to detect the motion of the Earth through the luminiferous aether. It could not find any trace of this kind of motion, suggesting, as a result, that it is impossible to detect one's presumed absolute motion, that is, motion with respect to the hypothesized luminiferous aether. The Michelson–Morley experiment said little about the speed of light relative to the light's source and observer's velocity, as both the source and observer in this experiment were traveling at the same velocity together in space.

The refractive index of a material indicates how much slower the speed of light is in that medium than in a vacuum. The slower speed of light in materials can cause refraction, as demonstrated by this prism (in the case of a prism splitting white light into a spectrum of colours, the refraction is known as dispersion).

In passing through materials, the observed speed of light can differ from c, as a result of the time lag between the polarization response of the medium and the incident light. The ratio of c to the phase velocity of light in the material is called the refractive index. The speed of light in air is only slightly less than *c*. Denser media, such as water and glass, can slow light much more, to fractions such as $^3D_4$ and $^2D_3$ of *c*. Through diamond, light is much slower—only about 124,000 km/s, less than ^1D $_2$of *c*. This reduction in speed is also responsible for bending of light at an interface between two materials with different indices, a phenomenon known as refraction.

Since the speed of light in a material depends on the refractive index, and the refractive index may depend on the frequency of the light, light at different frequencies can travel at different speeds through the same material. This effect is called dispersion.

Classically, considering electromagnetic radiation to be a wave, the charges of each atom (primarily the electrons) interact with the electric and magnetic fields of the radiation, slowing its progress.

Slow Light

Certain materials have an exceptionally high group index and a correspondingly low group velocity for light waves. In 1999, a team of scientists led by Lene Hau were able to slow the speed of a light pulse to about 17 m/s; in 2001, they were able to momentarily stop a beam.

In 2003, Mikhail Lukin, with scientists at Harvard University and the Lebedev Institute in Moscow, succeeded in completely halting light by directing it into a Bose–Einstein condensate of the element rubidium, the atoms of which, in Lukin's words, behaved "like tiny mirrors" due to an interference pattern in two "control" beams.

Faster-than-light Observations and Experiments

It is generally considered that it is impossible for any information or matter to travel faster than *c*, because it would travel backwards in time relative to some observers. However, there are many physical situations in which speeds greater than *c* are encountered.

Some of these situations involve entities that actually travel faster than *c* in a particular reference frame but none involves either matter, energy, or information traveling faster than light.

It is possible for the "group velocity" of light to exceed *c* and in an experiment in 2000 laser beams traveled for extremely short distances through caesium atoms with a group velocity of 300 times *c*. It is not, however, possible to use this technique to transfer information faster than *c* since the velocity of information transfer depends on the front velocity, which is always less than *c*.

Exceeding the group velocity of light in this manner is comparable to exceeding the speed of sound by arranging people distantly spaced in a line, and asking them all to shout "I'm here!", one after another with short intervals, each one timing it by looking at their own wristwatch so they don't have to wait until they hear the previous person shouting. Another example can be seen when watching ocean waves washing up on shore. With a narrow enough angle between the wave and the shoreline, the breakers travel along the waves' length much faster than the waves' movement inland.

If a laser is swept across a distant object, the spot of light can easily be made to move at a speed greater than *c*. Similarly, a shadow projected onto a distant object can be made to move faster than *c*. In neither case does any matter or information travel faster than light.

Quantum Mechanics

In quantum mechanics, certain quantum effects may be transmitted at speeds greater than *c*. For example, the quantum states of two particles can be entangled. Until the particles are observed, they exist in a superposition of two quantum states. If the particles are separated and one of them is observed to determine its quantum state then the quantum state of the second particle is determined automatically and faster than a light signal could travel between the two particles.

However, it is impossible to control which quantum state the first particle will take on when it is observed, so no information can be transmitted in this manner.

Closing speeds and proper speeds are examples of calculated speeds that may have value in excess of *c* but that do not represent the speed of an object as measured in a single inertial frame.

So-called superluminal motion is seen in certain astronomical objects, such as the jets of radio galaxies and quasars. However, these jets are not moving at speeds in excess of the speed of light: the apparent superluminal motion is a projection effect caused by objects moving near the speed of light and at a small angle to the line of sight.

It is possible for shock waves to be formed with electromagnetic radiation. If a charged particle travels through an insulating medium faster than the speed of light in that medium then radiation is emitted which is analogous to a sonic boom and is known as Èerenkov radiation.

Until relatively recent times, the speed of light was largely a matter of conjecture. Empedocles maintained that light was something in motion, and therefore there had to be some time elapsed in traveling. Aristotle said that, on the contrary, "light is due to the presence of something, but it is not a movement". Furthermore, if light had a finite speed, it would have to be very great; Aristotle asserted "the strain upon our powers of belief is too great" to believe this. The opposite view was argued by some, notably Roger Bacon. Euclid proposed the emission theory of vision, (also advanced by Ptolemy) where light was emitted from the eye, instead of entering the eye from another source. Using this theory, Heron of Alexandria advanced the argument that the speed of light must be infinite, since distant objects such as stars appear immediately upon opening the eyes.

Early Muslim philosophers initially agreed with the Aristotelian view of the speed of light being infinite. In 1021, however, the Iraqi physicist, Ibn al-Haytham (Alhazen), published the *Book of Optics*, in which he used experiments to support the intromission theory of vision, where light moves from an object into the eye, making use of instruments such as the camera obscura. This led to Alhazen proposing that light must therefore have a finite speed, and that the speed of light is variable, with its speed decreasing in denser bodies. He argued that light is a "substantial matter", the propagation of which requires time "even if this is hidden to our senses". This debate continued in Europe and the Middle East throughout the Middle Ages.

In the 11th century, Abû Rayhân al-Bîrûnî agreed that light has a finite speed and observed that the speed of light is much faster than the speed of sound. In the 1270s, Witelo considered the possibility of light traveling at infinite speed in a vacuum but slowing down in denser bodies. A comment on a verse in the *Rigveda* by the 14th century Indian scholar Sayana may be interpreted as suggesting an estimate for the speed of light that is in good agreement with its actual speed. In 1574, the Ottoman astronomer and physicist Taqi al-Din agreed with Alhazen that the speed of light is constant, but variable in denser bodies, and suggested that it would take a long time for light from the stars which are millions of kilometres away to reach the Earth.

In the early 17th century, Johannes Kepler believed that the speed of light was infinite since empty space presents no obstacle to it. Francis Bacon argued that the speed of light was not necessarily infinite, since something can travel too fast to be perceived. René Descartes argued that if the speed of light were finite, the Sun, Earth, and Moon would be noticeably out of alignment during a lunar eclipse. Since such misalignment had not been observed, Descartes concluded the speed of light was infinite. Descartes speculated that if the speed of light was found to be finite, his whole system of philosophy might be demolished.

Isaac Beeckman proposed an experiment (1629) in which a person would observe the flash of a cannon reflecting off a mirror about one mile (1.6 km) away. Galileo Galilei proposed an experiment (1638), with an apparent claim to having performed it some years earlier, to measure the speed of light by observing the delay between uncovering a lantern and its perception some distance away. He concluded that the speed of light is ten times faster than the speed of sound (in reality, light is around a million times faster than sound).

This experiment was carried out by the Accademia del Cimento of Florence in 1667, with the lanterns separated by about one mile (1.6 km). No delay was observed. Robert Hooke explained the negative results as Galileo had by pointing out that such observations did not establish the infinite speed of light, but only that the speed must be very great.

Astronomical Techniques

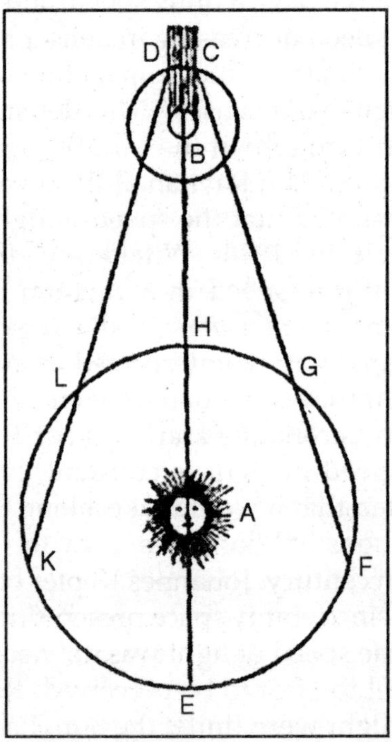

Fig. Rømer's Observations of the Occultations of Io from Earth.

The first quantitative estimate of the speed of light was made in 1676 by Ole Christensen Rømer, who was studying the motions of Jupiter's moon, Io, with a telescope. It is possible to time the orbital revolution of Io because it enters and exits Jupiter's shadow at regular intervals (at C or D).

Rømer observed that Io revolved around Jupiter once every 42.5 hours when Earth was closest to Jupiter (at H).

He also observed that, as Earth and Jupiter moved apart (as from L to K), Io's exit from the shadow would begin progressively later than predicted. It was clear that these exit "signals" took longer to reach Earth, as Earth and Jupiter moved further apart. This was as a result of the extra time it took for light to cross the extra distance between the planets, time which had accumulated in the interval between one signal and the next.

The opposite is the case when they are approaching (as from F to G). Rømer observed 40 orbits of Io when Earth was approaching Jupiter to be 22 minutes shorter than 40 orbits of Io when Earth was moving away from Jupiter. On the basis of those observations, Rømer concluded that it took light 22 minutes to cross the distance the Earth traversed in 80 orbits of Io. That corresponds to a ratio between the speed of light of the speed with which Earth orbits the sun of

$$80 \times \frac{42.5 \text{ hours}}{22 \text{ minutes}} \approx 9{,}300.$$

In comparison the modern value is about 10,100.

Around the same time, the astronomical unit was estimated to be about 140 million kilometres. The astronomical unit and Rømer's time estimate were combined by Christiaan Huygens, who estimated the speed of light to be 1,000 Earth diameters per minute, based on having misinterpreted Rømer's value of 22 minutes to mean the time it would take light to cross the diameter of the orbit of the Earth. This is about 220,000 kilometres per second (136,000 miles per second), 26% lower than the currently accepted value, but still very much faster than any physical phenomenon then known.

Isaac Newton also accepted the finite speed. In his 1704 book *Opticks* he reports the value of 16.6 Earth diameters per second (210,000 kilometres per second, 30% less than the actual value), which it seems he inferred for himself (whether from Rømer's data, or otherwise, is not known). The same effect was subsequently observed by Rømer for a "spot" rotating with the surface of Jupiter. And later observations also showed the effect with the three other Galilean moons, where it was more difficult to observe, thus laying to rest some further objections that had been raised.

Even if, by these observations, the finite speed of light may not have been established to everyone's satisfaction (notably Jean-Dominique Cassini's), after the observations of James Bradley (1728), the hypothesis of infinite speed was considered discredited. Bradley deduced that starlight falling on the Earth should appear to come from a slight angle, which could be calculated by comparing the speed of the Earth in its orbit to the speed of light. This "aberration of light", as it is called, was observed to be about 1/200 of a degree. Bradley calculated the speed of light as about 298,000 kilometres per second (185,000 miles per second). This is only slightly less than the currently accepted value (less than one percent). The aberration effect has been studied extensively over the succeeding centuries, notably by Friedrich Georg Wilhelm Struve and de:Magnus Nyrén.

The first successful measurement of the speed of light using an earthbound apparatus was carried out by Hippolyte Fizeau in 1849. (This measures the speed of light in air, which is slower than the speed of light in vacuum by a factor of the refractive index of air, about 1.0003.) Fizeau's experiment was conceptually similar to those proposed by Beeckman and Galileo. A beam of light was directed at a mirror several thousand metres away.

On the way from the source to the mirror, the beam passed through a rotating cog wheel. At a certain rate of rotation, the beam could pass through one gap on the way out and another on the way back. If *á* is the angle between two consecutive openings and *d* the distance between the toothed wheel and the mirror, then the tooth wheel must rotate with the angular speed (ω):

$$\omega = \frac{\alpha c}{2d}$$

in order for the light to pass through. Fizeau chose $d = 8$ km.

But at slightly higher or lower rates, the beam would strike a tooth and not pass through the wheel. Knowing the distance to the mirror, the number of teeth on the wheel, and the rate of rotation, the speed of light could be calculated. Fizeau reported the speed of light as 313,000 kilometres per second. Fizeau's method was later refined by Marie Alfred Cornu (1872) and Joseph Perrotin (1900).

Leon Foucault improved on Fizeau's method by replacing the cogwheel with a rotating mirror. Foucault's estimate, published in 1862, was 298,000 kilometres per second.

Foucault's method was also used by Simon Newcomb and Albert A. Michelson. Michelson began his lengthy career by replicating and improving on Foucault's method. If α is the angle between the normals to two consecutive facets and *d* the distance between the light source and the mirror, then the mirror must rotate with the angular speed (ω):

$$\omega = \frac{\alpha c}{2d}$$

in order for the light to pass through.

After the work of James Clerk Maxwell, it was believed that light travelled at a constant speed relative to the "luminiferous aether", the medium that was then thought to be necessary for the transmission of light. This speed was determined by the aether and its permittivity and permeability.

In 1887, the physicists Albert Michelson and Edward Morley performed the influential Michelson–Morley experiment to measure the velocity of the Earth through the aether. As shown in the diagram of a Michelson interferometer, a half-silvered mirror was used to split a beam of monochromatic light into two beams traveling at right angles to one another.

After leaving the splitter, each beam was reflected back and forth between mirrors several times (the same number for each beam to give a long but equal path length; the actual Michelson-Morley experiment used more mirrors than shown) then recombined to produce a pattern of constructive and destructive interference.

Any slight change in speed of light along one arm of the interferometer compared with its speed along the other arm (because the apparatus was moving with the Earth through the proposed "aether") would then be observed as a change in the pattern of interference. In the event, the experiment gave a null result.

Ernst Mach was among the first physicists to suggest that the experiment amounted to a disproof of the aether theory. Developments in theoretical physics had already begun to provide an alternative theory, Fitzgerald-Lorentz contraction, which explained the null result of the experiment.

It is uncertain whether Albert Einstein knew the results of the Michelson-Morley experiment, but the null result of the experiment greatly assisted the acceptance of his theory of relativity. The constant speed of light is one of the fundamental postulates (together with causality and the equivalence of inertial frames) of special relativity.

In 1926, Michelson used a rotating prism to measure the time it took light to make a round trip from Mount Wilson to Mount San Antonio in California, a distance of about 22 miles (36 km) each way. The precise measurements yielded a speed of 186,285 miles per second (299,796 kilometres per second).

During World War II, the development of the cavity resonance wavemeter for use in radar, together with precision timing methods, opened the way to laboratory-based measurements of the speed of light. In 1946, Louis Essen in collaboration with A.C. Gordon-Smith used a microwave cavity of precisely known dimensions to establish the frequency for a variety of normal modes of microwaves—which, in common with all electromagnetic radiation, travels at the speed of light in vacuum.

As the wavelength of the modes was known from the geometry of the cavity and from electromagnetic theory, knowledge of the associated frequencies enabled a calculation of the speed of light.

Their result, 299,792 ± 3 km/s, was substantially greater than those found by optical techniques, and prompted much controversy. However, by 1950 repeated measurements by Essen established a result of 299,792.5 ± 1 km/s; this became the value adopted by the 12th General Assembly of the Radio-Scientific Union in 1957. Most subsequent measurements have been consistent with this value.

With modern electronics (and most particularly the availability of oscilloscopes with time resolutions in the sub-nanosecond regime) the speed of light can now be directly measured by timing the delay of a light pulse from a laser or a LED in reflecting from a mirror, and this kind of experiment is now routine in undergraduate physics laboratories.

The metre is the length of the path travelled by light in vacuum during a time interval of 1/299,792,458 of a second. The second is the duration of 9,192,631,770 periods of the radiation corresponding to the transition between the two hyperfine levels of the ground state of the caesium-133 atom.

The consequence of this definition is that no experimental measurement could change the fact that the speed of light is exactly 299,792,458 metres per second.

A precise experimental measurement of the speed of light could, however, refine or alter the length of a metre.

GENERAL FIELD EQUATIONS

As simple and satisfying as Coulomb's equation may be, it is not entirely correct in the context of classical electromagnetism. Problems arise because changes in charge distributions require a non-zero amount of time to be "felt" elsewhere(required by special relativity).

Disturbances of the electric field due to a charge propagate at the speed of light.

For the fields of general charge distributions, the retarded potentials can be computed and differentiated accordingly to yield Jefimenko's Equations.

Retarded potentials can also be derived for point charges, and the equations are known as the Liénard-Wiechert potentials. These can then be differentiated accordingly to obtain the complete field equations for a moving point particle.

QUANTUM ELECTRODYNAMICS

Quantum electrodynamics(QED) is a relativistic quantum field theory of electrodynamics. QED was developed by a number of physicists, beginning in the late 1920s.

It basically describes how light and matter interacts. More specifically it deals with the interactions between electrons, positrons and photons. QED mathematically describes all phenomena involving electrically charged particles interacting by means of exchange of photons.

It has been called "the jewel of physics" for its extremely accurate predictions of quantities like the anomalous magnetic moment of the electron, and the Lamb shift of the energy levels of hydrogen.

In technical terms, QED can be described as a perturbation theory of the electromagnetic quantum vacuum. The history of quantum mechanics as this interlaces with history of quantum chemistry began essentially with the 1838 discovery of cathode rays by Michael Faraday, during the 1859-1860 winter statement of the black body radiation problem by Gustav Kirchhoff, the 1877 suggestion by Ludwig Boltzmann that the energy states of a physical system could be discrete, and the 1900 quantum hypothesis by Max Planck that any energy radiating atomic system can theoretically be divided into a number of discrete 'energy elements' *å*(epsilon) such that each of these energy elements is proportional to the frequency *í* with which they each individually radiate energy.

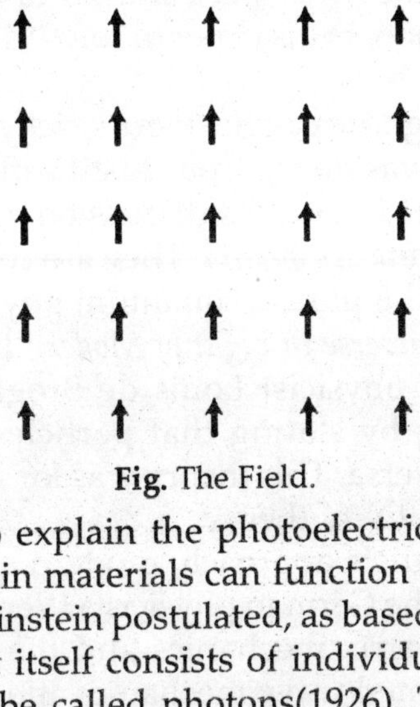

Fig. The Field.

Then, in 1905, to explain the photoelectric effect(1839), i.e. that shining light on certain materials can function to eject electrons from the material, Albert Einstein postulated, as based on Planck's quantum hypothesis, that light itself consists of individual quantum particles, which later came to be called photons(1926). The phrase "quantum mechanics" was first used in Max Born's 1924 paper "Zur

Quantenmechanik". In the years to follow, this theoretical basis slowly began to be applied to chemical structure, reactivity, and bonding.

In short, in 1900, German physicist Max Planck introduced the idea that energy is quantized, in order to derive a formula for the observed frequency dependence of the energy emitted by a black body. In 1905, Einstein explained the photoelectric effect by postulating that light, or more specifically all electromagnetic radiation, can be divided into a finite number of "energy quanta" that are localized points in space. From the introduction section of his March 1905 quantum paper, "On a heuristic viewpoint concerning the emission and transformation of light", Einstein states:

"According to the assumption to be contemplated here, when a light ray is spreading from a point, the energy is not distributed continuously over ever-increasing spaces, but consists of a finite number of *energy quanta* that are localized in points in space, move without dividing, and can be absorbed or generated only as a whole."

This statement has been called the most revolutionary sentence written by a physicist of the twentieth century. These *energy quanta* later came to be called "photons", a term introduced by Gilbert N. Lewis in 1926. The idea that each photon had to consist of energy in terms of quanta was a remarkable achievement as it effectively removed the possibility of black body radiation attaining infinite energy if it were to be explained in terms of wave forms only.

In 1913, Bohr explained the spectral lines of the hydrogen atom, again by using quantization, in his paper of July 1913 *On the Constitution of Atoms and Molecules*.

These theories, though successful, were strictly phenomenological: during this time, there was no rigorous justification for quantization aside, perhaps, for Henri Poincaré's discussion of Planck's theory in his 1912 paper *Sur la théorie des quanta*. They are collectively known as the *old quantum theory*. The phrase "quantum physics" was first used in Johnston's *Planck's Universe in Light of Modern Physics*(1931).

In 1924, the French physicist Louis de Broglie put forward his theory of matter waves by stating that particles can exhibit wave characteristics and vice versa. This theory was for a single particle and derived from special relativity theory.

Building on de Broglie's approach, modern quantum mechanics was born in 1925, when the German physicists Werner Heisenberg and Max Born developed matrix mechanics and the Austrian physicist Erwin Schrödinger invented wave mechanics and the non-relativistic Schrödinger equation as an approximation to the generalised case of de Broglie's theory. Schrödinger subsequently showed that the two approaches were equivalent.

Heisenberg formulated his uncertainty principle in 1927, and the Copenhagen interpretation started to take shape at about the same time. Starting around 1927, Paul Dirac began the process of unifying quantum mechanics with special relativity by proposing the Dirac equation for the electron.

The Dirac equation achieves the relativistic description of the wavefunction of an electron that Schrödinger failed to obtain. It predicts electron spin and led Dirac to predict the existence of the positron. He also pioneered the use of operator theory, including the influential bra-ket notation, as described in his famous 1930 textbook.

During the same period, Hungarian polymath John von Neumann formulated the rigorous mathematical basis for quantum mechanics as the theory of linear operators on Hilbert spaces, as described in his likewise famous 1932 textbook. These, like many other works from the founding period still stand, and remain widely used. The field of quantum chemistry was pioneered by physicists Walter Heitler and Fritz London, who published a study of the covalent bond of the hydrogen molecule in 1927. Quantum chemistry was subsequently developed by a large number of workers, including the American theoretical chemist Linus Pauling at Cal Tech, and John C. Slater into various theories such as Molecular Orbital Theory or Valence Theory.

Beginning in 1927, attempts were made to apply quantum mechanics to fields rather than single particles, resulting in what are known as quantum field theories. Early workers in this area included P.A.M. Dirac, W. Pauli, V. Weisskopf, and P. Jordan. This area of research culminated in the formulation of quantum electrodynamics by R.P. Feynman, F. Dyson, J. Schwinger, and S.I. Tomonaga during the 1940s. Quantum electrodynamics is a quantum theory of electrons, positrons, and the electromagnetic field, and served as a role model for subsequent quantum field theories. The theory of quantum chromodynamics was formulated beginning in the early 1960s.

The theory as we know it today was formulated by Politzer, Gross and Wilczek in 1975. Building on pioneering work by Schwinger, Higgs and Goldstone, the physicists Glashow, Weinberg and Salam independently showed how the weak nuclear force and quantum electrodynamics could be merged into a single electroweak force, for which they received the 1979 Nobel Prize in Physics.

The word 'quantum' is Latin, meaning "how much"(neut. sing. of quantus "how great"). The word 'electrodynamics' was coined by André-Marie Ampère in 1822.

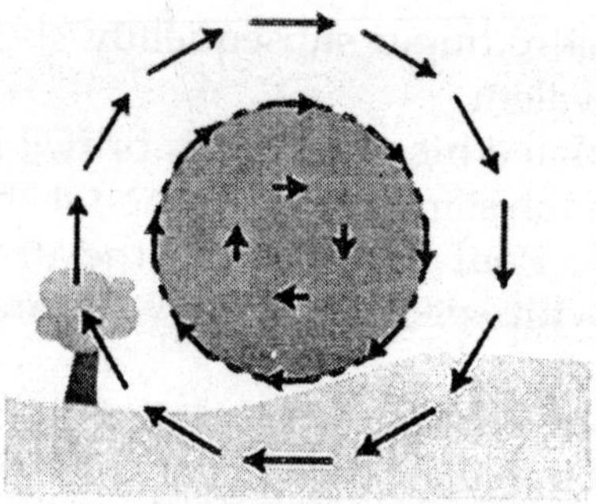

Fig. An Ampèrian Surface Superimposed on the Landscape.

The word 'quantum', as used in physics, i.e. with reference to the notion of count, was first used by Max Planck, in 1900 and reinforced by Einstein in 1905 with his use of the term *light quanta*. Quantum theory began in 1900, when Max Planck assumed that energy is quantized in order to derive a formula predicting the observed frequency dependence of the energy emitted by a black body.

This dependence is completely at variance with classical physics. In 1905, Einstein explained the photoelectric effect by postulating that light energy comes in quanta later called photons. In 1913, Bohr invoked quantization in his proposed explanation of the spectral lines of the hydrogen atom.

In 1924, Louis de Broglie proposed a quantum theory of the wave-like nature of subatomic particles. The phrase "quantum physics" was first employed in Johnston's *Planck's Universe in Light of Modern Physics*. These theories, while they fit the experimental facts to some extent, were strictly phenomenological: they provided no rigorous justification for the quantization they employed.

Modern quantum mechanics was born in 1925 with Werner Heisenberg's matrix mechanics and Erwin Schrödinger's wave mechanics and the Schrödinger equation, which was a non-relativistic generalization of de Broglie's(1925) relativistic approach.

Schrödinger subsequently showed that these two approaches were equivalent. In 1927, Heisenberg formulated his uncertainty principle, and the Copenhagen interpretation of quantum mechanics began to take shape. Around this time, Paul Dirac, in work culminating in his 1930 monograph finally joined quantum mechanics and special relativity, pioneered the use of operator theory, and devised the braket notation widely used since. In 1932, John von Neumann formulated the rigorous mathematical basis for quantum mechanics as the theory of linear operators on Hilbert spaces. This and other work from the founding period remains valid and widely used.

Quantum chemistry began with Walter Heitler and Fritz London's 1927 quantum account of the covalent bond of the hydrogen molecule. Linus Pauling and others contributed to the subsequent development of quantum chemistry.

The application of quantum mechanics to fields rather than single particles, resulting in what are known as quantum field theories, began in 1927. Early contributors included Dirac, Wolfgang Pauli, Weisskopf, and Jordan.

This line of research culminated in the 1940s in the quantum electrodynamics(QED) of Richard Feynman, Freeman Dyson, Julian Schwinger, and Sin-Itiro Tomonaga, for which Feynman, Schwinger and Tomonaga received the 1965 Nobel Prize in Physics. QED, a quantum theory of electrons, positrons, and the electromagnetic field, was the first satisfactory quantum description of a physical field and of the creation and annihilation of quantum particles.

QED involves a covariant and gauge invariant prescription for the calculation of observable quantities. Feynman's mathematical technique, based on his diagrams, initially seemed very different from the field-theoretic, operator-based approach of Schwinger and Tomonaga, but Freeman Dyson later showed that the two approaches were equivalent.

The renormalization procedure for eliminating the awkward infinite predictions of quantum field theory was first implemented in QED. Even though renormalization works very well in practice, Feynman was never entirely comfortable with its mathematical validity, even referring to renormalization as a "shell game" and "hocus pocus".

QED has served as the model and template for all subsequent quantum field theories. One such subsequent theory is quantum chromodynamics, which began in the early 1960s and attained its present form in the 1975 work by H.

David Politzer, Sidney Coleman, David Gross and Frank Wilczek. Building on the pioneering work of Schwinger, Peter Higgs, Goldstone, and others, Sheldon Glashow, Steven Weinberg and Abdus Salam independently showed how the weak nuclear force and quantum electrodynamics could be merged into a single electroweak force.

PHYSICAL INTERPRETATION OF QED

In classical optics, light travels over all allowed paths and their interference results in Fermat's principle. Similarly, in QED, light(or any other particle like an electron or a proton) passes over every possible path allowed by apertures or lenses.

The observer(at a particular location) simply detects the mathematical result of all wave functions added up, as a sum of all line integrals. For other interpretations, paths are viewed as non physical, mathematical constructs that are equivalent to other, possibly infinite, sets of mathematical expansions. According to QED, light can go slower or faster than c, but will travel at velocity c on average.

Physically, QED describes charged particles(and their antiparticles) interacting with each other by the exchange of photons. The magnitude of these interactions can be computed using perturbation theory; these rather complex formulas have a remarkable pictorial representation as Feynman diagrams.

QED was the theory to which Feynman diagrams were first applied. These diagrams were invented on the basis of Lagrangian mechanics. Using a Feynman diagram, one decides every possible path between the start and end points. Each path is assigned a complex-valued probability amplitude, and the actual amplitude we observe is the sum of all amplitudes over all possible paths.

The paths with stationary phase contribute most(due to lack of destructive interference with some neighboring counter-phase paths) – this results in the stationary classical path between the two points. QED doesn't predict what will happen in an experiment, but it can predict the *probability* of what will happen in an experiment, which is how it is experimentally verified. Predictions of QED agree with experiments to an extremely high degree of accuracy: currently about 10 (and limited by experimental errors).

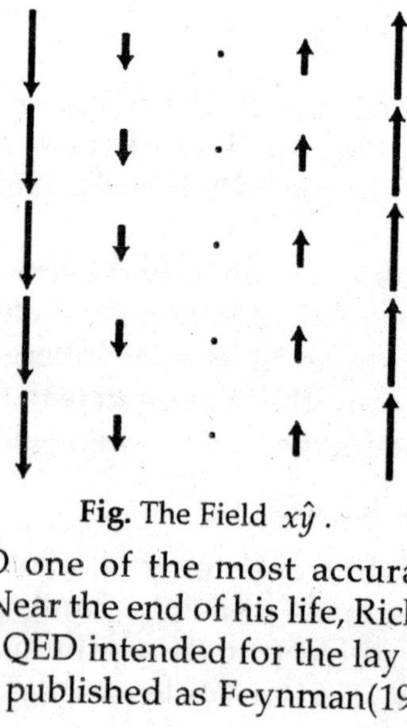

Fig. The Field $x\hat{y}$.

This makes QED one of the most accurate physical theories constructed thus far. Near the end of his life, Richard P. Feynman gave a series of lectures on QED intended for the lay public. These lectures were transcribed and published as Feynman(1985), *QED: The strange theory of light and matter*, a classic non-mathematical exposition of QED from the point of view articulated above.

MATHEMATICS

Mathematically, QED is an abelian gauge theory with the symmetry group U(1). The gauge field, which mediates the interaction between the charged spin–1/2 fields, is the electromagnetic field.

In Pictures

The part of the Lagrangian containing the electromagnetic field tensor describes the free evolution of the electromagnetic field, whereas the Dirac-like equation with the gauge covariant derivative describes the free evolution of the electron and positron fields as well as their interaction with the electromagnetic field.

SPEED OF LIGHT

"Lightspeed" redirects here. For other uses, see Lightspeed (disambiguation). The speed of light in the vacuum of free space is an important physical constant usually denoted by the symbol c0 or simply c. The metre is defined such that the speed of light in free space is exactly 299,792,458 metres per second (m/s).

The speed of light is of fundamental importance in physics. It is the speed of not just visible light, but of all electromagnetic radiation, as well as gravitational waves and anything having zero rest mass. In Einstein's theory of relativity the speed of light plays the crucial role of a conversion factor between space and time within spacetime. This theory together with the principle of causality requires that no matter or information can travel faster than the speed of light.

The speed of light is so great that for many purposes it can be regarded as infinite. However, where long distances or accurate time measurements are involved, the finite speed of light can be important. For example, in the Global Positioning System (GPS), a GPS receiver measures its distance to satellites based on how long it takes for a radio signal to arrive from the satellite. In astronomy, distances are often measured in light-years, the distance light travels in a year (around ten trillion kilometers).

The speed of light when it passes through a transparent or translucent material medium, like glass or air, is less than its speed in a vacuum. The speed is controlled by the refractive index of the medium. In specially-prepared media, the speed can be tiny, or even zero.

The speed of light in vacuum is now viewed as a fundamental

physical constant. This postulate, together with the principle of relativity that all inertial frames are equivalent, forms the basis of Einstein's theory of special relativity.

Experimental evidence has shown that the speed of light is independent of the motion of the source. It has also been confirmed experimentally that the two-way speed of light (for example from a source, to a mirror, and back again) is constant. It is not, however, possible to measure the one-way speed of light (for example from a source to a distant detector) without some convention as to how clocks at the source and receiver should be synchronized. Einstein (who was aware of this fact) postulated that the speed of light should be taken as constant in all cases, one-way and two-way.

An observer moving with respect to a collection of light sources would find that light from the sources ahead would be blueshifted while light from those behind was redshifted.

Use of the Symbol ?c' for the Speed of Light

The symbol c for 'constant' or the Latin celeritas ("swiftness") is used for the speed of light in free space, and in this article c is used exclusively this way. However, some authors use c for the speed of light in material media. To avoid confusion, and for consistency with other constants of free space such as μ_0, ε_0 and Z_0, international bodies such as the International Bureau of Weights and Measures(BIPM) recommend using c_0 for the speed of light in free space.

In branches of physics in which the speed of light plays an important part, such as in relativity, it is common to use a system of units known as "natural units" in which c is 1; thus no symbol for the speed of light is required.

Causality and Information Transfer

According to the theory of special relativity, causality would be violated if information could travel faster than c in one reference frame. In some other reference frames, the information would be received before it had been sent, so the "effect" could be observed before the "cause". Such a violation of causality has never been recorded.

To put it another way, information propagates to and from a point from regions defined by a light cone. The interval AB in the diagram to the right is "time-like" (that is, there is a frame of reference in which event A and event B occur at the same location in space, separated only by their occurring at different times, and if A precedes B in that frame then A precedes B in all frames: there is no frame of reference in which event A and event B occur simultaneously). Thus, it is hypothetically possible for matter (or information) to travel from A to B, so there can be a causal relationship (with A the "cause" and B the "effect").

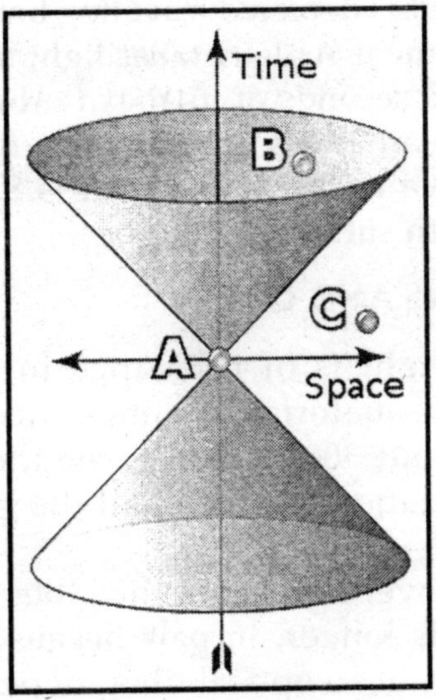

Fig. A Light Cone Defines Locations that are in Causal contact and those that are not.

On the other hand, the interval AC in the diagram to the right is "space-like" (that is, there is a frame of reference in which event A and event C occur simultaneously, separated only in space; see simultaneity). However, there are also frames in which A precedes C or in which C precedes A. Barring some way of traveling faster than light, it is not possible for any matter (or information) to travel from A to C or from C to A. Thus there is no causal connection between A and C.

Speed of Light in Astronomy

The speed of light is particularly important in astronomy. Due to the vast distances involved it can take a very long time for light to travel from its source to Earth. For example, it takes 13 billion years for light to travel to Earth from the faraway galaxies viewed in the Hubble Ultra Deep Field images. Those photographs, taken today, capture images of the galaxies as they appeared 13 billion years ago (near the beginning of the universe). The fact that farther-away objects appear younger (due to the finite speed of light) is crucial in astronomy, allowing astronomers to infer the evolution of stars, galaxies, and the universe itself.

Astronomical distances are sometimes measured in light-years, the distance light travels in one year. A light year is around 9 trillion km, 6 trillion miles, or 0.3 parsecs. The closest star to Earth (besides the sun) is around 4.2 light years away.

The beam of light is depicted traveling between the Earth and the Moon in the same time it actually takes light to scale the real distance between them: 1.255 seconds at its mean orbital distance (surface to surface). The light beam helps provide the sense of scale of the Earth-Moon system relative to the Sun, which is 8.28 light-minutes away (photosphere to Earth surface).

COMMUNICATIONS AND GPS

The speed of light is of relevance to communications. For example, given the equatorial circumference of the Earth is about 40,075 km and c about 300,000 km/s, the theoretical shortest time for a piece of information to travel half the globe along the surface is 0.066838 s.

When light is traveling around the globe in an optical fiber, the actual transit time is longer, in part because the speed of light is slower by about 35% in an optical fiber, depending on its refractive index n, v = c/n. Furthermore, straight lines rarely occur in global communications situations, and delays are created when the signal passes through an electronic switch or signal regenerator. A typical time as of 2004 for a U.S. to Australia or Japan computer-to-computer ping is 0.18 s. The speed of light additionally affects wireless communications design.

Another consequence of the finite speed of light is that communications between the Earth and spacecraft are not instantaneous. There is a brief delay from the source to the receiver, which becomes more noticeable as distances increase. This delay was significant for communications between ground control and Apollo 8 when it became the first spacecraft to orbit the Moon: for every question, the ground control station had to wait at least three seconds for the answer to arrive. The communications delay between Earth and Mars is almost ten minutes. As a consequence of this, if a robot on the surface of Mars were to encounter a problem, its human controllers would not be aware of it until ten minutes later; it would then take at least a further ten minutes for instructions to travel from Earth to Mars.

This effect forms the basis of the Global Positioning System (GPS) and similar navigation systems. A position on Earth can be determined by means of the delays in radio signals received from a number of satellites, each carrying a very accurate atomic clock, and very carefully synchronized. To work properly, this method requires that (among many other effects) the relative motion of satellite and receiver be taken into effect, which was how (on an interplanetary scale) the finite speed of light was originally discovered (see the following section). The speed of light can also be of concern over very short distances. In supercomputers, the speed of light imposes a limit on how quickly data can be sent between processors.

If a processor operates at 1 GHz, a signal can only travel a maximum of 300 mm (about one foot) in a single cycle. Processors must therefore be placed close to each other to minimize communication latencies, which can cause difficulty with cooling. If clock frequencies continue to increase, the speed of light will eventually become a limiting factor for the internal design of single chips.

Constant Velocity from all Inertial Reference Frames

Most individuals are accustomed to the addition rule of velocities: if two cars approach each other from opposite directions, each traveling at a speed of 50 km/h, relative to the road surface, one expects that each car will measure the other as approaching at a combined speed of 50 + 50 = 100 km/h to a very high degree of accuracy. However, as speeds increase this rule becomes less accurate. Two spaceships approaching each other, each traveling at 90% the speed of light relative to some third observer, would not measure each other as approaching at 90% + 90% = 180% the speed of light; instead they each measure the other as approaching at slightly less than 99.5% the speed of light. This last result is given by the Einstein velocity-addition formula:

$$u = \frac{v_1 + v_2}{1 + (v_1 \cdot v_2)/c^2}$$

where v_1 and v_2 are the velocities of the spaceships as measured by the third observer, and u is the measured velocity of either space ship as observed by the other. This reduces to {{{1}}} for sufficiently small values of v_1 and v_2 (such as those typically encountered in common daily experiences), as the term $(v_1 \cdot v_2)/c_2$ approaches zero, reducing the denominator to 1.

If one of the velocities for the above formula (or both) are c, the final result is c, as is expected if the speed of light is the same in all reference frames. Another important result is that this formula always returns a value which is less than c whenever v_1 and v_2 are less than c: this shows that no acceleration in any frame of reference can cause one to exceed the speed of light with respect to another observer. Thus c acts as a speed limit for all objects with respect to all other objects in special relativity.

Luminiferous Aether (Discredited)

Before the advent of special relativity, it was believed that light traveled through a medium called the luminiferous aether. Maxwell's equations predict a given speed of light, in much the same way as is the speed of sound in air.

The speed of sound in air is relative to the movement of the air itself, and the speed of sound in air with respect to an observer may be changed if the observer is moving with respect to the air (or vice versa). The speed of light was believed to be relative to a medium of transmission for light that acted as air does for the transmission of sound-the luminiferous aether.

The Michelson-Morley experiment, arguably the most famous and useful null-result experiment in the history of physics, was designed to detect the motion of the Earth through the luminiferous aether. It could not find any trace of this kind of motion, suggesting, as a result, that it is impossible to detect one's presumed absolute motion, that is, motion with respect to the hypothesized luminiferous aether. The Michelson-Morley experiment said little about the speed of light relative to the light's source and observer's velocity, as both the source and observer in this experiment were traveling at the same velocity together in space.

The refractive index of a material indicates how much slower the speed of light is in that medium than in a vacuum. The slower speed of light in materials can cause refraction, as demonstrated by this prism (in the case of a prism splitting white light into a spectrum of colours, the refraction is known as dispersion).

In passing through materials, the observed speed of light can differ from c, as a result of the time lag between the polarization response of the medium and the incident light.

The ratio of c to the phase velocity of light in the material is called the refractive index.

The speed of light in air is only slightly less than c. Denser media, such as water and glass, can slow light much more, to fractions such as and of c. Through diamond, light is much slower-only about 124,000 km/s, less than c. This reduction in speed is also responsible for bending of light at an interface between two materials with different indices, a phenomenon known as refraction.

Since the speed of light in a material depends on the refractive index, and the refractive index may depend on the frequency of the light, light at different frequencies can travel at different speeds through the same material. This effect is called dispersion.

Classically, considering electromagnetic radiation to be a wave, the charges of each atom (primarily the electrons) interact with the electric and magnetic fields of the radiation, slowing its progress.

Slow Light

Certain materials have an exceptionally high group index and a correspondingly low group velocity for light waves. In 1999, a team of scientists led by Lene Hau were able to slow the speed of a light pulse to about 17 m/s; in 2001, they were able to momentarily stop a beam.

In 2003, Mikhail Lukin, with scientists at Harvard University and the Lebedev Institute in Moscow, succeeded in completely halting light by directing it into a Bose-Einstein condensate of the element rubidium, the atoms of which, in Lukin's words, behaved "like tiny mirrors" due to an interference pattern in two "control" beams.

Faster-than-Light Observations and Experiments

It is generally considered that it is impossible for any information or matter to travel faster than c, because it would travel backwards in time relative to some observers. However, there are many physical situations in which speeds greater than c are encountered.

Some of these situations involve entities that actually travel faster than c in a particular reference frame but none involves either matter, energy, or information traveling faster than light.

It is possible for the "group velocity" of light to exceed c and in an experiment in 2000 laser beams traveled for extremely short distances through caesium atoms with a group velocity of 300 times c. It is not, however, possible to use this technique to transfer information faster than c since the velocity of information transfer depends on the front velocity, which is always less than c.

Exceeding the group velocity of light in this manner is comparable to exceeding the speed of sound by arranging people distantly spaced in a line, and asking them all to shout "I'm here!", one after another with short intervals, each one timing it by looking at their own wristwatch so they don't have to wait until they hear the previous person shouting. Another example can be seen when watching ocean waves washing up on shore. With a narrow enough angle between the wave and the shoreline, the breakers travel along the waves' length much faster than the waves' movement inland.

If a laser is swept across a distant object, the spot of light can easily be made to move at a speed greater than c. Similarly, a shadow projected onto a distant object can be made to move faster than c. In neither case does any matter or information travel faster than light.

Quantum Mechanics

In quantum mechanics, certain quantum effects may be transmitted at speeds greater than c. For example, the quantum states of two particles can be entangled. Until the particles are observed, they exist in a superposition of two quantum states. If the particles are separated and one of them is observed to determine its quantum state then the quantum state of the second particle is determined automatically and faster than a light signal could travel between the two particles. However, it is impossible to control which quantum state the first particle will take on when it is observed, so no information can be transmitted in this manner.

Closing speeds and proper speeds are examples of calculated speeds that may have value in excess of c but that do not represent the speed of an object as measured in a single inertial frame.

So-called superluminal motion is seen in certain astronomical objects, such as the jets of radio galaxies and quasars. However, these jets are not moving at speeds in excess of the speed of light: the apparent superluminal motion is a projection effect caused by objects moving near the speed of light and at a small angle to the line of sight.

It is possible for shock waves to be formed with electromagnetic radiation. If a charged particle travels through an insulating medium faster than the speed of light in that medium then radiation is emitted which is analogous to a sonic boom and is known as ?erenkov radiation.

Until relatively recent times, the speed of light was largely a matter of conjecture. Empedocles maintained that light was something in motion, and therefore there had to be some time elapsed in traveling. Aristotle said that, on the contrary, "light is due to the presence of something, but it is not a movement". Furthermore, if light had a finite speed, it would have to be very great; Aristotle asserted "the strain upon our powers of belief is too great" to believe this. The opposite view was argued by some, notably Roger Bacon.

Euclid proposed the emission theory of vision, (also advanced by Ptolemy) where light was emitted from the eye, instead of entering the eye from another source. Using this theory, Heron of Alexandria advanced the argument that the speed of light must be infinite, since distant objects such as stars appear immediately upon opening the eyes.

Early Muslim philosophers initially agreed with the Aristotelian view of the speed of light being infinite. In 1021, however, the Iraqi physicist, Ibn al-Haytham (Alhazen), published the Book of Optics, in which he used experiments to support the intromission theory of vision, where light moves from an object into the eye, making use of instruments such as the camera obscura. This led to Alhazen proposing that light must therefore have a finite speed, and that the speed of light is variable, with its speed decreasing in denser bodies. He argued that light is a "substantial matter", the propagation of which requires time "even if this is hidden to our senses". This debate continued in Europe and the Middle East throughout the Middle Ages.

In the 11th century, Abu Rayhan al-Biruni agreed that light has a finite speed and observed that the speed of light is much faster than the speed of sound. In the 1270s, Witelo considered the possibility of light traveling at infinite speed in a vacuum but slowing down in denser bodies. A comment on a verse in the Rigveda by the 14th century Indian scholar Sayana may be interpreted as suggesting an estimate for the speed of light that is in good agreement with its actual speed. In 1574, the Ottoman astronomer and physicist Taqi al-Din agreed with Alhazen that the speed of light is constant, but variable in denser bodies, and suggested that it would take a long time for light from the stars which are millions of kilometres away to reach the Earth.

In the early 17th century, Johannes Kepler believed that the speed of light was infinite since empty space presents no obstacle to it. Francis Bacon argued that the speed of light was not necessarily infinite, since something can travel too fast to be perceived. René Descartes argued that if the speed of light were finite, the Sun, Earth, and Moon would be noticeably out of alignment during a lunar eclipse. Since such misalignment had not been observed, Descartes concluded the speed of light was infinite. Descartes speculated that if the speed of light was found to be finite, his whole system of philosophy might be demolished.

Isaac Beeckman proposed an experiment (1629) in which a person would observe the flash of a cannon reflecting off a mirror about one mile (1.6 km) away. Galileo Galilei proposed an experiment (1638), with an apparent claim to having performed it some years earlier, to measure the speed of light by observing the delay between uncovering a lantern and its perception some distance away. He concluded that the speed of light is ten times faster than the speed of sound (in reality, light is around a million times faster than sound).

This experiment was carried out by the Accademia del Cimento of Florence in 1667, with the lanterns separated by about one mile (1.6 km). No delay was observed. Robert Hooke explained the negative results as Galileo had by pointing out that such observations did not establish the infinite speed of light, but only that the speed must be very great.

Astronomical Techniques

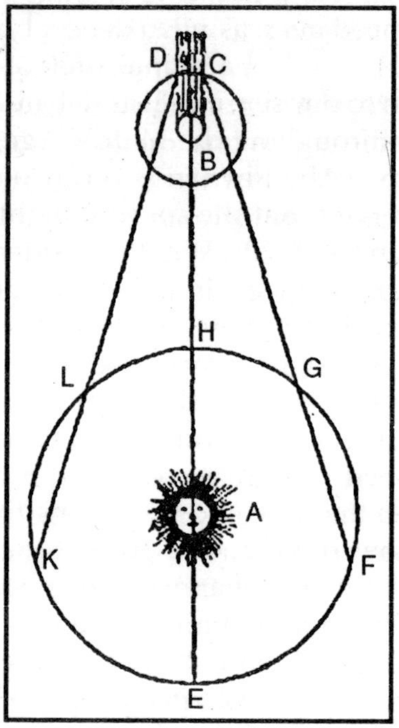

Fig. Rømer's observations of the occultations of Io from Earth.

The first quantitative estimate of the speed of light was made in 1676 by Ole Christensen Rømer, who was studying the motions of Jupiter's moon, Io, with a telescope. It is possible to time the orbital revolution of Io because it enters and exits Jupiter's shadow at regular intervals (at C or D). Rømer observed that Io revolved around Jupiter once every 42.5 hours when Earth was closest to Jupiter (at H).

He also observed that, as Earth and Jupiter moved apart (as from L to K), Io's exit from the shadow would begin progressively later than predicted. It was clear that these exit "signals" took longer to reach Earth, as Earth and Jupiter moved further apart. This was as a result of the extra time it took for light to cross the extra distance between the planets, time which had accumulated in the interval between one signal and the next.

The opposite is the case when they are approaching (as from F to G). Rømer observed 40 orbits of Io when Earth was approaching Jupiter to be 22 minutes shorter than 40 orbits of Io when Earth was moving away from Jupiter. On the basis of those observations, Rømer concluded that it took light 22 minutes to cross the distance the Earth traversed in 80 orbits of Io. That corresponds to a ratio between the speed of light of the speed with which Earth orbits the sun of

$$80 \times \frac{42.5 \text{ hours}}{22 \text{ minutes}} \approx 9{,}300.$$

In comparison the modern value is about 10,100.

Around the same time, the astronomical unit was estimated to be about 140 million kilometres. The astronomical unit and Rømer's time estimate were combined by Christiaan Huygens, who estimated the speed of light to be 1,000 Earth diameters per minute, based on having misinterpreted Rømer's value of 22 minutes to mean the time it would take light to cross the diameter of the orbit of the Earth. This is about 220,000 kilometres per second (136,000 miles per second), 26% lower than the currently accepted value, but still very much faster than any physical phenomenon then known.

Isaac Newton also accepted the finite speed. In his 1704 book Opticks he reports the value of 16.6 Earth diameters per second (210,000 kilometres per second, 30% less than the actual value), which it seems he inferred for himself (whether from Rømer's data, or otherwise, is not known). The same effect was subsequently observed by Rømer for a "spot" rotating with the surface of Jupiter. And later observations also showed the effect with the three other Galilean moons, where it was more difficult to observe, thus laying to rest some further objections that had been raised.

Even if, by these observations, the finite speed of light may not have been established to everyone's satisfaction (notably Jean-Dominique Cassini's), after the observations of James Bradley (1728), the hypothesis of infinite speed was considered discredited. Bradley deduced that starlight falling on the Earth should appear to come from a slight angle, which could be calculated by comparing the speed of the Earth in its orbit to the speed of light. This "aberration of light", as it is called, was observed to be about 1/200 of a degree. Bradley calculated the speed of light as about 298,000 kilometres per second (185,000 miles per second). This is only slightly less than the currently accepted value (less than one percent). The aberration effect has been studied extensively over the succeeding centuries, notably by Friedrich Georg Wilhelm Struve and de:Magnus Nyrén.

The first successful measurement of the speed of light using an earthbound apparatus was carried out by Hippolyte Fizeau in 1849.

(This measures the speed of light in air, which is slower than the speed of light in vacuum by a factor of the refractive index of air, about 1.0003.) Fizeau's experiment was conceptually similar to those proposed by Beeckman and Galileo. A beam of light was directed at a mirror several thousand metres away. On the way from the source to the mirror, the beam passed through a rotating cog wheel. At a certain rate of rotation, the beam could pass through one gap on the way out and another on the way back. If α is the angle between two consecutive openings and d the distance between the toothed wheel and the mirror, then the tooth wheel must rotate with the angular speed (ω):

$$\omega = \frac{\alpha c}{2d}$$

in order for the light to pass through. Fizeau chose d = 8 km.

But at slightly higher or lower rates, the beam would strike a tooth and not pass through the wheel. Knowing the distance to the mirror, the number of teeth on the wheel, and the rate of rotation, the speed of light could be calculated. Fizeau reported the speed of light as 313,000 kilometres per second. Fizeau's method was later refined by Marie Alfred Cornu (1872) and Joseph Perrotin (1900).

Leon Foucault improved on Fizeau's method by replacing the cogwheel with a rotating mirror. Foucault's estimate, published in 1862, was 298,000 kilometres per second. Foucault's method was also used by Simon Newcomb and Albert A. Michelson. Michelson began his lengthy career by replicating and improving on Foucault's method. If ? is the angle between the normals to two consecutive facets and d the distance between the light source and the mirror, then the mirror must rotate with the angular speed(ω):

$$\omega = \frac{\alpha c}{2d}$$

in order for the light to pass through

After the work of James Clerk Maxwell, it was believed that light travelled at a constant speed relative to the "luminiferous aether", the medium that was then thought to be necessary for the transmission of light. This speed was determined by the aether and its permittivity and permeability.

In 1887, the physicists Albert Michelson and Edward Morley performed the influential Michelson-Morley experiment to measure the velocity of the Earth through the aether. As shown in the diagram of a Michelson interferometer, a half-silvered mirror was used to split a beam of monochromatic light into two beams traveling at right angles to one another.

After leaving the splitter, each beam was reflected back and forth between mirrors several times (the same number for each beam to give a long but equal path length; the actual Michelson-Morley experiment used more mirrors than shown) then recombined to produce a pattern of constructive and destructive interference. Any slight change in speed of light along one arm of the interferometer compared with its speed along the other arm (because the apparatus was moving with the Earth through the proposed "aether") would then be observed as a change in the pattern of interference. In the event, the experiment gave a null result.

Ernst Mach was among the first physicists to suggest that the experiment amounted to a disproof of the aether theory. Developments in theoretical physics had already begun to provide an alternative theory, Fitzgerald-Lorentz contraction, which explained the null result of the experiment.

It is uncertain whether Albert Einstein knew the results of the Michelson-Morley experiment, but the null result of the experiment greatly assisted the acceptance of his theory of relativity. The constant speed of light is one of the fundamental postulates (together with causality and the equivalence of inertial frames) of special relativity.

In 1926, Michelson used a rotating prism to measure the time it took light to make a round trip from Mount Wilson to Mount San Antonio in California, a distance of about 22 miles (36 km) each way. The precise measurements yielded a speed of 186,285 miles per second (299,796 kilometres per second).

During World War II, the development of the cavity resonance wavemeter for use in radar, together with precision timing methods, opened the way to laboratory-based measurements of the speed of light. In 1946, Louis Essen in collaboration with A.C. Gordon-Smith used a microwave cavity of precisely known dimensions to establish the frequency for a variety of normal modes of microwaves-which, in common with all electromagnetic radiation, travels at the speed of light in vacuum.

As the wavelength of the modes was known from the geometry of the cavity and from electromagnetic theory, knowledge of the associated frequencies enabled a calculation of the speed of light. Their result, 299,792 ± 3km/s, was substantially greater than those found by optical techniques, and prompted much controversy. However, by 1950 repeated measurements by Essen established a result of 299,792.5 ± 1 km/s; this became the value adopted by the 12th General Assembly of the Radio-Scientific Union in 1957. Most subsequent measurements have been consistent with this value.

With modern electronics (and most particularly the availability of oscilloscopes with time resolutions in the sub-nanosecond regime) the speed of light can now be directly measured by timing the delay of a light pulse from a laser or a LED in reflecting from a mirror, and this kind of experiment is now routine in undergraduate physics laboratories.

The metre is the length of the path travelled by light in vacuum during a time interval of 1/299,792,458 of a second. The second is the duration of 9,192,631,770 periods of the radiation corresponding to the transition between the two hyperfine levels of the ground state of the caesium-133 atom.

The consequence of this definition is that no experimental measurement could change the fact that the speed of light is exactly 299,792,458 metres per second. A precise experimental measurement of the speed of light could, however, refine or alter the length of a metre.

Index

A

Alnico 51
Ampere Model 41
Antiferromagnetism 46, 59
Antimatter 21
Atomic dipoles 83

B

Benjamin 11, 104, 105, 107
Bonding 256
Broglie 33, 149, 256, 258

C

Ceramic 50, 51, 52
Circular Loop 10, 17
Classical History 100
Concentric 10, 68
Conductor Problem 77
Continuous Field 165, 169, 170, 174, 179, 182, 183, 191
Conventions 41, 90
Counterrotating 10
Creationism 22, 23

D

Dark Matter 150, 174, 175, 176, 179, 180, 182, 185, 186, 187, 188
Demagnetization 49
Derivation 209, 222, 261
Destructionism 22, 23
Diamagnetism 44, 58, 116
Dipole Operator 83
Dirac String 91, 92
Direction of Force 65
Distant Field 6, 9
Domains 45, 46, 74, 170, 192, 197

E

Electric Dipole 9, 10, 79, 80, 82, 85, 86
Electric Field 1, 5, 28, 32, 56, 60, 64, 66, 67, 68, 70, 71, 72, 76, 77, 78, 81, 82, 86, 166, 169, 174, 230, 232, 235, 239, 240, 241, 254
Electric Force 60, 67, 78, 122, 166, 168
Electric Machine 104, 105, 106, 110, 111, 113, 115, 119
Electrical Currents 64, 68
Electrodynamics 2, 11, 13, 36, 38, 67, 92, 113, 131, 132, 144, 219, 227, 228, 229, 230, 232, 233, 255, 257, 259
Electromagnetic Force 3, 12, 16, 49
Electromagnets 39, 53, 54, 69

F

Faraday 3, 38, 108, 111, 112, 114, 115, 116, 122, 125, 164, 165, 166, 167, 174, 175, 180, 182, 233, 255
Ferrite 50, 51
Ferromagnetism 44, 46, 47, 59, 98
Flexible 51, 52

G

Galaxies 151, 152, 153, 155, 157, 158, 159, 161, 162, 163, 172, 221

Grand Unified 70, 88, 89, 92, 95, 96, 99
Gravitational Lensing 151, 152, 162

H

Hall Effect 73
Heavy Minerals 48
Henry 3, 14, 108, 112, 115, 116, 118, 121, 123, 126, 231, 133
Hypothetical 70, 71, 88, 91, 124, 223, 224

I

Identical Particles 19
Industrial Revolution 127, 166
Injection 51
Interpretation 23, 34, 74, 91, 149, 165, 177, 186, 187, 188, 189, 194, 203, 209, 218, 257, 258, 259

L

Leyden jar 105, 106, 107, 108, 111, 118, 122
Linear Materials 12
Lorentz 12, 13, 44, 65, 71, 73, 77, 131, 132, 194, 225, 226, 230

M

Magnet 37, 38, 39, 40, 41, 42, 45, 46, 47, 48, 49, 50, 51, 52, 53, 54, 55, 56, 58, 59, 60, 61, 62, 63, 64, 66, 67, 69, 70, 72, 74, 76, 77, 78, 80, 81, 87, 88, 89, 98, 103, 115, 116, 118, 126
Magnetic Behaviours 42
Magnetic Cipole 66, 67, 76
Magnetic Dipole 10, 39, 41, 53, 59, 64, 66, 67, 69, 70, 74, 79, 80, 81, 84, 85, 86, 87, 88
Magnetic Field 1, 2, 3, 5, 6, 7, 8, 10, 12, 13, 14, 16, 28, 32, 37, 38, 39, 40, 41, 42, 43, 44, 46, 48, 49, 50, 53, 54, 55, 56, 57, 58, 60, 61, 62, 63, 64, 65, 66, 67, 68, 69, 70, 71, 72, 73, 74, 75, 76, 77, 78, 79, 80, 81, 84, 86, 87, 88, 89, 90, 94, 119, 134, 166, 231, 232, 235, 239, 240, 241
Magnetic Force 1, 55
Magnetic Hand 48
Magnetic Materials 38, 39, 43, 50, 52, 54, 74, 75, 76
Magnetic Monopole 40, 70, 88, 90, 91, 96, 97, 98
Magnetic Pole 40, 41, 54, 64, 87
Magnetization 39, 40, 41, 42, 43, 44, 47, 49, 52, 53, 54, 55, 74, 117, 166
Magnetometer 73
Magnets 1, 37, 38, 39, 40, 41, 42, 45, 47, 48, 49, 50, 51, 52, 53, 54, 55, 60, 61, 63, 66, 77, 78, 87, 89, 98, 102
Magnitude 39, 40, 56, 63, 70, 71, 73, 77, 79, 80, 84, 134, 140, 146, 149, 186, 193, 195, 203, 204, 226, 234, 260
Maxwell 3, 12, 15, 38, 72, 73, 76, 121, 122, 123, 135, 166, 167, 194, 207, 230, 231
Mechanical 5, 19, 20, 25, 26, 44, 47, 48, 51, 79, 81, 83, 101, 104, 144, 215, 216, 225, 226, 227, 228, 230
Metallic elements 50
Middle Ages 102
Midplane 6, 7, 9, 10
Molded 51
Molecular Dipoles 81, 82
Monopole 40, 70, 81, 88, 89, 90, 91, 92, 94, 96, 97, 98
Moving Charge 2, 64, 65, 67, 68, 77, 79

N

Nano-Structured 52
Non-Electrics 104, 105
Non-Uniform 64, 66, 67, 81, 186, 191, 192

P

Paramagnetism 42, 43, 58, 116
Particle Model 182, 234
Permanent Magnets 39, 45, 50, 52, 54, 60, 61, 89

Phenomenon 1, 2, 20, 23, 28, 59, 78, 79, 99, 100, 110, 111, 114, 122, 140, 141, 147, 153, 155, 191, 194, 195, 225, 226, 235, 238
Photoelectric Effect 13, 132, 255, 256, 258
Physical Dipoles 81
Physical Interpretation 74, 259
Poincare 13, 131, 132, 226
Point Dipoles 81
Popular Culture 97
Probability 17, 18, 21, 97, 155, 157, 169, 189, 260
Propagation 120, 121, 122, 134, 136, 152, 153, 167, 170, 171, 194, 195, 196, 231, 235, 240, 241

Q

Quantization 16, 28, 88, 90, 91, 93, 99, 227, 256, 258

R

Radiation 3, 13, 15, 23, 27, 28, 29, 30, 35, 73, 86, 94, 136, 137, 139, 140, 141, 142, 144, 147, 149, 150, 167, 192, 194, 196, 197, 223, 225, 226, 227, 230, 231, 232, 233, 234, 235, 236, 237, 238, 255, 256
Radio waves 3, 73, 167, 230, 231, 234, 236, 237, 238
Radioactivity 15, 19, 20
Rare Earth 51, 50, 52
Relativity 2, 12, 13, 15, 16, 28, 36, 38, 59, 60, 77, 78, 131, 132, 140, 151, 152, 166, 170, 173, 179, 180, 181, 182, 183, 184, 185, 188, 189, 190, 191, 192, 194, 196, 197, 198, 201, 213, 227, 228, 235, 254, 256, 257, 258
Renaissance 102
Resinous 105, 107
Rutherford Atom 29, 30, 143, 144

S

Single-Chain 52
Single-Molecule 52
Spectroscopy 141, 142, 237
Spectrum 27, 137, 138, 139, 141, 142, 145, 150, 162, 177, 192, 230, 231, 233, 234, 235, 236, 237, 238
Steady Current 68
String theory 89, 94
Superposition 6, 87, 135, 136, 175, 189, 232, 234

T

Tesla 14, 53, 54, 71, 76, 77, 84, 121, 123, 124, 126, 127
Ticonal 51
Topological 91, 93, 94, 95, 96
Torque 40, 58, 64, 66, 71, 86, 88
Transition 95, 171, 191, 199, 222

U

Unified Theorics 70, 88, 89, 92, 95, 99

V

Vector Potential 72, 85, 90, 91, 230
Visualizing 62, 63
Vitreous 105, 107

W

Waves 3, 15, 28, 33, 37, 52, 60, 70, 73, 121, 123, 134, 136, 137, 149, 167, 192, 193, 230, 231, 232, 233, 234, 236, 237, 238, 240, 241, 256
Weak Lensing 158, 159, 160, 161, 162